Saunders Mac Lane

Saunders Mac Lane
A Mathematical Autobiography

Saunders Mac Lane
University of Chicago, Emeritus

A K Peters
Wellesley, Massachusetts

Editorial, Sales, and Customer Service Office

A K Peters, Ltd.
888 Worcester Street, Suite 230
Wellesley, MA 02482
www.akpeters.com

Copyright © 2005 by A K Peters, Ltd.

All rights reserved. No part of the material protected by this copyright notice may be reproduced or utilized in any form, electronic or mechanical, including photocopying, recording, or by any information storage and retrieval system, without written permission from the copyright owner.

Library of Congress Cataloging-in-Publication Data
Mac Lane, Saunders, 1909-
   Saunders Mac Lane : a mathematical autobiography / Saunders Mac Lane.
     p. cm.
   Includes index.
   ISBN 1-56881-150-0 (alk. paper)
    1. Mac Lane, Saunders, 1909- 2. Mathematicians—United States—Biography. I. Title.

QA29.M24A3 2004
510'.92–dc22
[B]

                                                          2004044638

Developmental editing by Janet Beissinger.

Printed in Canada
09 08 07 06 05   10 9 8 7 6 5 4 3 2 1

Contents

Preface	ix
Acknowledgments	xv
Part One: Early Years	**1**
Chapter 1: Heresy	3
Chapter 2: High School	13
Chapter 3: Undergraduate at Yale	21
Chapter 4: Universal Knowledge and New Knowledge	31
Chapter 5: The University of Chicago, 1930	35
Chapter 6: Germany 1931-33	43
Part Two: First Teaching	**63**
Chapter 7: Yale and Harvard	65
Chapter 8: Cornell and Chicago	75
Chapter 9: Surveying Modern Algebra	81
Chapter 10: Algebraic Functions	83
Chapter 11: First Graduate Students	87
Part Three: Collaborative Research	**91**
Chapter 12: Crossed Product Algebras and Group Extension	93
Chapter 13: Eilenberg Enters	99
Chapter 14: Naturality	105
Part Four: The War Years	**111**
Chapter 15: Much More Applied Math	113
Chapter 16: Cynthia Enters	117
Chapter 17: War Research: Roll, Pitch, and Yaw	119

Contents

Part Five: Eilenberg and Mac Lane — 123
Chapter 18: Cautiously Publishing Category Theory — 125
Chapter 19: The Cohomology of Groups — 127
Chapter 20: Eilenberg-Mac Lane Spaces — 131

Part Six: Harvard Years — 133
Chapter 21: Professor at Harvard — 135
Chapter 22: University Presidents — 141
Chapter 23: Jib and Mainsail — 143
Chapter 24: Dorothy — 147
Chapter 25: Have Guggenheim, Will Travel — 155

Part Seven: Chicago in the Fifties — 165
Chapter 26: Return to Chicago — 167
Chapter 27: The Stone Age at Chicago — 171
Chapter 28: The Stone Age Comes to an End — 177
Chapter 29: Hutchins and the University — 183
Chapter 30: The College Mathematics Staff — 189
Chapter 31: Univeral Algebra and Think Tanks — 191

Part Eight: Mathematical Developments — 195
Chapter 32: Mathematical Organizations — 197
Chapter 33: Bourbaki-the Legend — 201
Chapter 34: The New Math — 205
Chapter 35: Categories Expand — 209
Chapter 36: The Grand Tour of Europe, 1954 — 211
Chapter 37: Paris and Cartan. 1955-56 — 219

Part Nine: National Academy of Science — 225
Chapter 38: Membership in the National Academy of Sciences — 227
Chapter 39: The National Research Council — 229
Chapter 40: The Academy Proceedings — 233

Part Ten: The Sixties and Beyond — 235
Chapter 41: Homological Algebra — 237

Contents

Chapter 42: Categories: La Jolla & Columbia — 239
Chapter 43: Geometrical Mechanics — 243
Chapter 44: Outdoors on the Indiana Dunes — 245
Chapter 45: Categories at Work — 249

Part Eleven: National Science Policy — **253**
Chapter 46: As President of the AMS — 255
Chapter 47: Academy Reports — 259
Chapter 48: George Kistiakowski — 265
Chapter 49: Report Review — 267
Chapter 50: The National Science Board — 279
Chapter 51: Science Policy — 287

Part Twelve: Travels — **291**
Chapter 52: Visits to China — 293
Chapter 53: Anniversary at the Dunes, 1983 — 299
Chapter 54: Dorothy's Delights — 303

Part Thirteen: Advising — **307**
Chapter 55: Chicago Graduate Students — 309
Chapter 56: Friends and Mentors — 317
Chapter 57: Rating Research — 319
Chapter 58: The NAS Research Roundtable — 321

Part Fourteen: Later Developments — **323**
Chapter 59: The Philosophy of Mathematics — 325
Chapter 60: Second Marriage — 327
Chapter 61: International Category Conferences — 333

Part Fifteen: Contemplating — **337**
Chapter 62: Mathematics Departments — 339
Chapter 63: Collaborative Research — 345
Chapter 64: Career Choice: Inheritance of Precision? — 349
Index — 353

Preface

> No man could so stimulate others unless, alongside an incisive intellect, he was possessed of enthusiasm and warmth, a deep interest in his fellow man, and a sympathy the more real for being unsentimental. Those who proudly call themselves his friends know these things: others will infer them in reading [his works].
>
> —M. Kelly[1]

Saunders Mac Lane has been my teacher, mentor, and model almost from the beginning of my mathematical life. It is a relationship I've cherished. He has been for me a figure of great honesty and integrity, who worked hard to advance research and to serve the mathematical community. His belief in the good, the right, and the rational, his care for the essence of mathematical ideas, his powerful enthusiasm, and his essential optimism were and are deeply attractive to me.

Nearly everything about Saunders in action was colorful, starting with the red-and-green plaid sports coat (the Mac Lane tartan, of course) and red pants that he would wear for important occasions. Perhaps a few anecdotes and reflections from my experience of him over 40 years will help the reader appreciate this color.

First Encounter

I first met Mac Lane—in a sense I'll make precise—in 1963. He was one of the most important figures in the University of Chicago

---

[1] From *Saunders Mac Lane: Selected Papers*, edited by I. Kaplansky, New York, Springer-Verlag, 1979.

Preface

Mathematics Department, or indeed in American mathematics: his first student, Irving Kaplansky, was Chair of the department, and two other students were on the faculty—one, John Thompson, a Fields Medalist. Mac Lane was an inventor of group cohomology, a founder of homological algebra and category theory, and known for the Eilenberg-Mac Lane spaces in topology. He was past President of the Mathematical Association of America, and he would soon be Vice President of the National Academy of Sciences, member of the Board governing the National Science Foundation, and President of the American Mathematical Society, as well.

I knew none of this. I was sixteen, an early entrant to the University, an uneven student with a great enthusiasm for mathematics. It was the beginning of my second quarter, and I was scheduled to start a basic linear algebra class that morning. I happened to arrive a little early, settled down in the first row of the class, and sank peacefully into a daydream. Being so new, I wasn't surprised not to know the other students who settled in around me, and I didn't know the teacher that I'd have. In due course, Mac Lane walked in and began lecturing. His style was lively and colorful, and I was immediately interested—but almost at once aware that I'd made a big mistake: this was not an undergraduate linear algebra course but an advanced graduate course on category theory. I'd come an hour early.

I understood nothing whatever after a few moments, but was far too embarrassed to get up and leave—instead I sank into daydreams, glassy-eyed. Mac Lane, who prided himself on paying attention to his class, later told me he thought that he could always see who was following and who was not. In a moment like a thunderclap, I looked up from my seat and found him pointing directly at me from across the room. "You!" he said peremptorily, "you don't believe this proof, do you?" Belief and disbelief were equally beyond me; I sat petrified. He advanced toward me, and I don't know what I imagined—that he would pick me up by the scruff of my neck and throw me from the room? He stopped, turned back to the board, and proceeded to explain the proof to satisfy me. Of course, I still understood nothing—but I sat in rapt attention.

Preface

Fortunately the class ended soon, and as students asking questions surrounded him, it was easy for me to slip out. I didn't tell Saunders this story until many years afterwards, when I had the privilege of re-enacting it (from the other side) in a lecture at the conference in honor of his seventieth birthday. Needless to say, the event hadn't left a trace in his memory, though it remains sharp for me to this day.

Saunders and Tolerance

Saunders believes strongly in principles, in the rightness of right positions. I never once saw him personally intolerant, but he could sometimes be direct and candid to the point of offending. People whose judgment I respect have felt injured by what he said and sometimes by the bluntness of his expression. In some way perhaps he didn't appreciate the magnitude of his position in mathematics or the seriousness with which people took him. In a lesser personage some of his extreme positions might have been regarded as charming eccentricities. But given Saunders' stature, they could injure, and he might have been more cautious.

An event from late in Saunders' life may give a bit of the flavor. It was a special session run by him and Richard Askey at the Joint Mathematics Meeting in 1999, a session boldly entitled "Mathematics Education and Mistaken Philosophies of Mathematics." The audience was enormous. I found the title charming (and still find it so, even now as I become more involved with ideas in K-12 education), and I imagine that Saunders meant it to be controversial but playful. Predictably, it annoyed and needled some practitioners. Saunders began the session with introductory remarks that I found fascinating: he said that he now considered the extent of his own emphasis on category theory as a tool for learning and teaching mathematics to have been too extreme. This humbleness may have helped soften the critical tone of the session.

Saunders and Sammy

One of Saunders' great mathematical friendships and collaborations was with Samuel Eilenberg (widely known as Sammy, or even S^2P^2: "Smart Sammy the Polish Prodigy"). I got to see them in action together only once, at the AMS Summer Research Institute on Category Theory at Bowdoin College, in 1969. They had special status at this three-week conference, not only as the senior members but also as the very founders of the subject. So, when they began discussing its origins one evening after dinner, everyone gathered around to listen.

I dearly wish I could recall the substance of their debate, but I don't; only my sense of the contrast in the two men's styles stays with me. Sammy drew Saunders out and egged him on, always slightly evasive and mocking; Saunders, whose father and grandfather were Congregational ministers, seemed to feel that since his view was right, his view would prevail. Once he had stated it, all he could do was bang his fist. The devious and sophisticated European versus the innocent but honest American? That's how it seemed to me at the time (maybe I was a little innocent myself). A loyal student, I was rooting from the beginning for Saunders' point of view, but I came away feeling that he was trounced in the contest.

Being Saunders' Student

After I flirted a while with operator theory (Paul Halmos and Felix Browder were my teachers) and group theory (learned from Jon Alperin and Otto Kegel), it was finally time for me, by now a second-year graduate student, to settle on an area for a PhD thesis. I obsessed about how to make the choice. A close mathematical friend, Joe Neisendorfer, explained to me an algorithm: forget the topic, look around the faculty for the person you like the most. It didn't take me long to choose Saunders.

I wouldn't say I ever felt personal intimacy with Saunders, but he did go out of his way to make me and other students feel welcome in more than his office. Saunders and his late wife, Dorothy, had a small but comfortable cottage in the Indiana Dunes, a beautiful area on the shore of Lake Michigan about an hour south of Chicago, and they occasionally invited students to spend an afternoon there. Saunders was an enthusiastic sailor, and I can report, from a ride in a small sailboat on rough water, that he was ready to provide needed

Preface

instruction not only in mathematics but also on how to handle the absence of a toilet—or any privacy—in that difficult situation.

If you look at the list of Saunders' 39 students, you'll see that Irving Kaplansky, who worked on valuation theory of fields, came first; I'm near the end, with a thesis on noncommutative rings. Along the way are such people as John Thompson (finite groups), Anil Nerode (logic and computation), and Robert Szczarba (algebraic topology). How did this variety come about?

Perhaps the answer lies in Saunders' hospitality to these many ideas. He wanted to learn finite groups and taught a course on them. By the end of the course, he'd decided that he'd never really understand the subject, but in Thompson he found a fabulously strong student. Saunders might have tried to turn such a student toward interests close to his own, but I think he would not, on principle: he was happy to encourage his students to do what excited them.

Saunders has followed an interesting, curving trajectory through mathematics, from logic and foundations to field theory and the beginnings of homological algebra, through topology to category theory, with smaller diversions along the way into Hamiltonian mechanics, finite groups, and many other subjects. Perhaps his students, or many of them, could be described as coming off of the tangents to this path, a kind of developable surface reaching broadly across mathematics. Altogether, Saunders has more than 1,000 mathematical descendents listed on the Mathematics Genealogy Project.[2]

Some other aspects of Saunders are also reflected in his students: Saunders was always active on behalf of the community, whether as Chair working to build the department at the University of Chicago or, near the end of his career, as member of the National Science Board or as manager of the elaborate system of reports for the National Academy of Sciences. Many of his students and grand-students have followed him into this willingness for public service. When I was worrying about whether to move to my current position at MSRI, he was one of the first people I called on for advice and blessing, and he gave both.

---

[2] A service of the Department of Mathematics at North Dakota State University, available at http://www.genealogy.ams.org/.

Preface

Returning to the more fundamental matter of being Saunders' mathematical student: I tried for a while, dutifully, to find a thesis topic in category theory, Saunders' passion in that part of his life. But I failed; somehow, the things I read and learned in that domain just didn't inspire me. When I developed an interest instead in a problem on noncommutative rings posed by a visitor of Herstein, the young Chris Robson, Saunders could easily have washed his hands of the project. He did not: though it was far from his current area of interest, he welcomed what I had done, and painstakingly read draft after draft of my thesis.

Saunders' mode of instruction in thesis writing bears mention. I had written a couple of papers, jointly with Robson, of which my thesis results were partially an extract. Robson cared a lot about exposition, and so (learning from Saunders among others) did I. We'd gone through many drafts, and I thought the writing pretty polished. Saunders did not. He began at the beginning and worked his way through the thesis until he'd compiled a list of exactly 25 substantive suggestions. Then he stopped and returned the document to me for an overhaul. When I had finished making the corrections he'd flagged and all their analogues, I gave it back to him, eager to be done. But...after a week or so I got a second list of exactly 25 more suggestions. The third list was a bit shorter, and Saunders allowed the process to converge before I got too frustrated.

It must be clear by now: over these forty years I've learned many lessons from Saunders. I'm deeply grateful to him.

David Eisenbud,
Berkeley, California

Acknowledgments

When Saunders Mac Lane asked me, a few years ago, if I would publish his autobiography, I was honored and very pleased because I knew that he has been witness to so many interesting, diverse, and important events of which he had been an astute and formative observer. I first met Saunders when I joined Springer Verlag as the first in-house mathematics editor in 1964 and he was an author and one of the editors of the series Grundlehren der Mathematischen Wissenschaften. Over the years I received much valuable advice and shared many stories with Saunders, but when I read the first draft of his autobiography, I found so much that I had not known and insights that I am sure will be educational and inspirational to young mathematicians and colleagues of many years.

It is my duty and pleasure to thank those who have assisted, contributed to, and enhanced this manuscript; I do so also on behalf of Saunders whose health does not allow him to express all the thanks that are due at this time.

The first person I want to thank is Dr. Janet Beissinger, a mathematician at the University of Illinois at Chicago who answered our request to read and develop the draft of the manuscript that Saunders had mostly completed. Our first meeting with the author led to an extended series of regular meetings and conversations that gave Janet an understanding of the motivation for the project and Saunders' goals and led to her extensive development of the manuscript in close cooperation with Saunders. These meetings gave Saunders

Acknowledgments

the opportunity to test and expand his narrative with a sympathetic yet independently minded reader/listener and to approve her edits. We cannot thank Janet enough for her selfless service. She graciously acknowledges that the project expanded her perspective and taught her many things.

Thanks are due to Greta Schuessler, who was the Assistant Executive Officer of the Report Review Committee of the National Academy of Science during Saunders' tenure in that position. Her reading of and commenting on the relevant sections of the book were invaluable, yet all responsibility for the final manuscript rests with the author.

Gretchen Mac Lane, a professional editor in her own right, read the manuscript and amended and clarified memories while leaving the original narrative and voice intact; her younger sister, Cynthia Hay, added comments and corrections.

Osa Mac Lane encouraged the project and served as diplomatic guardian during the many sessions and conversations we all had with Saunders, assuring his patience with our repeated questions. She helped collect the wonderful pictures that are included in the book and, I am sure, was a strong support behind Saunders' efforts to collect and put down his memories.

Irving Kaplansky and David Eisenbud, two of Saunders' many students, encouraged the project from the very beginning and have been helpful along the way with advice and encouragement. Two other students of Saunders, Paul Palmquist and John MacDonald, have carefully read the manuscript and made suggestions that have been implemented.

My thanks extend to all of the above and to those who have helped without my awareness.

With this I turn the manuscript into the hands of the reading public and hope that it will be received with interest, pleasure, and appreciation for a life of creativity and service.

Klaus Peters
Publisher

Part One

Early Years

Chapter One
Heresy

Castle Duart rises grim and foreboding over the Straits of Mull in the Highlands of Scotland, the former center of the clan MacLean. The MacLeans conceded victory to the British in the last battle of the Forty-Five Rebellion, the Battle of Culloden and later suffered the fate of many other Highlanders who were evicted from their homes to make way for sheep farms during the "Highland Clearances" of the early 1800s. Thus, in the early 19th century, my MacLean ancestors sought new beginnings in western Pennsylvania and Ohio.

My grandfather, William Ward McLane—son of John and Julia McLane—was born on November 18, 1846, in Lewisville, Indiana County, Pennsylvania. William graduated from Blackburn College in 1871, and from the Western Theological Seminary in 1874. He was then ordained to the Presbyterian ministry, and served as pastor of the Second Presbyterian Church in Stuebenville, Ohio until 1883. That year, William was officially charged with heresy for preaching on the writings of Charles Darwin. Evolution was widely unaccepted among the faithful of the time, and at the annual meeting of the Presbyterian Synod it became clear to my grandfather that he would be convicted of his charge.

He therefore resigned his position from the church in 1884 and set off with his family to New Haven, where he became pastor of the College Street Congregational Church. At this time, William had three sons, John F. and Paul from his first marriage, and baby Donald

Early Years

Reverend William Ward MacLane

(my father), the eldest son of his second marriage to Frances Robinson. Frances was an enthusiastic poet and descendant of William Bradford, who had sailed on the Mayflower and had become one of the first governors of the Massachusetts Bay Colony.

Once settled in New Haven, William moved the church from its traditional location on the New Haven Green to a more outlying location and continued his scholarly work, eventually earning a Doctor of Divinity at Yale.

The family lived near the Yale campus at 85 Howe Street for a considerable amount of time. The address is noteworthy because Josiah Willard Gibbs, the theoretical physicist and chemist who became one of the greatest American scientists of the 19th century, lived on the same block. The eminence of this neighbor undoubtedly contributes to the persistence of a family legend that Gibbs, while taking refuge from a thunderstorm in a doorway, came upon my grandfather and asked, "Young man, have not I seen you before?" Of course, Gibbs had seen my grandfather on many previous occasions, only without the dubious benefit of rain to inspire conversation.

After many years in New Haven, William was retired from his pastorate, probably because the congregants wanted a younger pastor. He moved to North Leominster, Massachusetts, and became pastor

Chapter One ~ Heresy

The Saunders family in Newport, RI, 1913 (top row from left to right: Donald, George holding Lois, Isabel holding Tom Jr., Tom Powell, Sr.; front row from left to right: Saunders, Winifred, Aretas, Priscilla, Dorothea)

of the Congregational Church. We will return to him there later in my story.

My father, Donald Bradford McLane, was born January 19, 1882, and was a year old on the journey from Stuebenville to New Haven when his parents fled their parish to avoid charges of heresy. In New Haven, he graduated from the well-known Hill House High School in 1899. He was a bright and well-liked member of his school; in his senior class book he wrote the essay covering junior year, and was voted the wittiest and most eccentric member of his class. He studied at Yale, where he graduated in 1903, and went on to the Union Theological Seminary in New York City. After his graduation there, he served for a year as an assistant pastor of the Church of the Sea and Land.

The next-door neighbor, George Aretas Saunders was a parishioner of his father's church in New Haven. He had moved from Newport, Rhode Island, where his father, Aretas Saunders, was a prominent dentist. He attended the Sheffield Scientific School at Yale as an engineering student, but gave up school to marry Isabel Andrews. They remained in New Haven and raised three children: Winifred,

Early Years

Saunders as a young child with his parents, Donald McLane and Winifred Saunders

Dorothea, and Aretas. Winifred, the eldest, was beautiful and talented. She graduated Phi Beta Kappa from Mount Holyoke College and briefly taught high-school English and mathematics. Though the Saunders and McLane families lived next door from one another, my father was shy about approaching Winifred. But we know he must have followed family advice that "faint heart never won fair lady," as he and Winifred were married in 1908.

After their marriage, Donald took up his first regular pastorate at the Congregational Church in Taftville, Connecticut. Taftville was a New England mill town, dominated by a cotton mill and surrounded by simple workers' houses.

I was born nearby in Norwich on August 4, 1909, christened Leslie Saunders MacLane. My father and his brothers had changed the spelling of the family name from MacLean to MacLane so as not to be considered Irish. It was my nurse who suggested the name Leslie, but a month later, my parents agreed that they didn't like the name. My father put his hand on my head, looked up to God, and said, "Leslie forget." I have gone by two last names ever since. The space in Mac Lane was added years later by me, when my first wife, Dorothy, found it difficult to type our name without a space.

Chapter One ~ Heresy

In Taftville we lived on top of a hill with a good view of the town, and life there remains idyllic in my memory—I was quite unaware of the life of the workers in the mill and their connection to my father's church. I remember vividly sliding down our hill with my father, and I subsequently loved slides and sliding forever. I still have a Flexible Flyer in my basement, and in Cambridge I once built a backyard snow slide, slanted 60 degrees, for my three-year-old daughter, Gretchen, who reminds me how I systematically iced the surface for a slick slide down.

In the back of the Taftville house there were extensive woods. My father took me on walks from an early age, and I fondly remember that he encouraged me to find my own way home. I developed a good sense of direction and later encouraged the same for my own daughters by pointing out "markers" and the direction of the sun on the walk out, letting the girls find the way back using these clues.

After four years in the mill town, my family moved to Boston, where my father became pastor of a Congregational church in Roxbury/Jamaica Plain. This stands in my memory as a great change, and many events impressed on me an alienation and anonymity of city life that contrasted with my experience in a small town. Our family walks now took place in a park around a lake instead of in the woods, and our backyard was paved over. My father had helped me construct a playhouse out of a wooden crate, but older boys in the neighborhood destroyed it. I also remember social tensions that were frightening and unfamiliar to me—there were race riots nearby that ended in stonings.

My family suffered a loss during this time—the birth of my sister Lois had happily completed the family, but she died of heart failure at the age of 4. My parents were devastated, and my father celebrated her brief life in poetry, a talent that he had inherited from his mother. The loss was difficult to accept, and my parents attempted to overcome it through the adoption of another daughter, but she didn't settle well into the family, and the adoption was not completed.

In the meantime, I did start attending school. There is little I recall. On one occasion, I did not return directly home from school, but wandered about instead. I was suitably lectured when I finally

Saunders and his sister Lois, soon before she died

arrived at home. I do specifically remember learning fractions in the third grade. My teacher used cut-up strips of paper to demonstrate their meaning, which impressed me. I like to think of that lesson in today's terminology—the activity would be called "hands-on learning," and the strips "manipulatives"—and know that my teacher used these "modern" methods in 1917.

Politics mattered then, and I closely followed my father's views. In 1916 he favored the reelection of President Wilson, who at that time had "kept us out of the war." I was impressed with this slogan and put on a campaign for Wilson. I apparently felt that at age 7, I had sound political judgment, and left VOTE FOR WILSON slips all around the neighborhood.

When the United States entered the war in 1917, political issues became more serious. My father maintained his pacifism, which resulted in tensions with many of his parishioners. Hence, we moved to North Wilbraham, Massachusetts, a small farm village outside Springfield, where my father became pastor of a combined church. Previously, there was a Methodist and Congregational church, perhaps too many for a small town, and so they united, but there was considerable doctrinal tension in the combined church.

Chapter One ~ Heresy

Saunders, age 5, and Lois, age 2, with their parents

    I was happy to be in a small country town again. There was a pond nearby, and a dump full of wonderful, old discarded objects. The parsonage had a big backyard with a fine sand pile where I played at soldiers. Our neighbor was a Civil War veteran, which I found very impressive. We had a garden in the yard that had more space than we needed for planting, and I took over one row and constructed a mock trench. I could not have understood the horrors of warfare, but the war, though taking place at a great distance, had influence.

    The grade school in Wilbraham was in a three-room building, though I remember using only two of them; one for grades one to five, and the other for grades six to eight. Many years later, I returned to Wilbraham to find that my old school had been preserved as an example of how education used to be. I recall little of the subject matter I learned there; life outside of school was much more exciting. I played with my friends at the dump, and in the bushes of the pond. We sometimes broke into an abandoned hardware store, and in the winter, I delighted in riding my sled down the town hills. Once, I even rode in an automobile, which was a complete wonder to me at the time. My brother Gerald was born at the hospital in Springfield in 1918, which finally filled the void in the family left by Lois's death.

    Well before completion of the expected six-year pastorate in Wilbraham, my family moved to a new parish. The reason for the move could well have been doctrinal disputes with Methodists, or possibly my father's part-time job. The Wilbraham Academy was a boarding school for boys just down the street, and the headmaster

Early Years

was a prominent Methodist parishioner. At the start of the school year, a faculty member abruptly resigned, and my father was recruited to fill his place as the Latin teacher at the Academy. My father knew many languages, including Latin; he once wrote an article about the Lord's Prayer in 27 languages. But teaching was a distraction for my father. He preferred to write sermons and visit his parishioners; he felt that visitations were an essential aspect of his ministry.

My father's new parish was in Utica, New York. Utica was a growing city that had expanded southward along Genesee Street. The Congregationalists set up a brand new church there, and my father was the founding pastor. This presented a challenge for him: building the church, recruiting new members, preaching sermons, and comforting parishioners were daunting responsibilities. Because the church was brand-new, there was no existing parsonage. My father had to buy one of the new jerry-built houses that had a then-considerable mortgage of $6,500. At this time I was only 12 years old, but I was concerned by the troubles of my family's finances.

The local grammar school was just around the corner from our house. I attended and found it very different from my previous two-room schoolhouse. At the end of the seventh grade, all students had to take the New York State Regents examination in arithmetic. The requirements involved quickly adding up long columns of four-digit numbers, a skill that my fellow students had practiced much more than me. It is my recollection that I flunked this exam, but apparently there were no resulting sanctions, as I went on to the eighth grade.

The city of Utica offered some rural aspects: a creek passed through the wrecked site of an amusement park nearby, continued through a stretch of woods, and then flowed under the main street by a culvert. Together with my friend Bill Schmidt we explored the culvert and, probably recklessly, waded through it under the street. We used abandoned lumber from the amusement park to build a splendid tree house eight feet up, which we accessed by ladder. We even included a fireplace in the tree house, though its design left something to be desired—we were evidently unaware that proper construction of a fireplace included a draft, and when we lighted our

Chapter One ❧ Heresy

Saunders with his brothers Gerald and David in a homebuilt cart, Utica, NY, 1922

fire, smoke promptly filled the tree house. We then built a second tree house, this time without a fireplace.

I engaged in other constructions; in the backyard of our house I put together a mock sailboat, and I also used an old set of wheels to build a cart in which my brother Gerald could ride—he did not seem to mind that I volunteered him for this activity. My family bought me a bicycle, which I used to earn some money by delivering baked goods from the home of Mrs. Crippen on the next block. My delivery was not, apparently, always careful; a customer complained that the meringue on her lemon meringue pie had suffered slippage. I tried to do better after that, as I was quite happy to be earning money.

In the winter, the town hills provided ample opportunity for sledding. My old Flexible Flyer had grown too small, and so I used the money from my delivery to buy a wonderful new sled—a Flexible Flyer Racer. Misfortune soon ensued. I took my new sled down a toboggan slide on a big hill, lost control, and badly bent a runner. Disappointed, I took it back to the department store, where, to my surprise (still today), they willingly replaced it with a new sled. I still have that fine Flexible Flyer Racer; it provides a small link to my boyhood.

I also joined a troop of Boy Scouts that met in a church downtown, where we enjoyed games such as shuffleboard. I took seriously to scouting, enough to spend my money on a uniform, which I proudly wore to the hospital where my mother had given birth to my youngest brother, David. As a Scout, I enjoyed two weeks of summer camp, though I did fail badly at some of the athletic tests.

I began high school in the center of the city, at the Utica Free Academy. Though I must have taken the usual courses for freshmen, I remember only algebra, which was taught by a veteran teacher who was also the football coach. And of algebra, I remember only my pleasure at learning that it was indeed possible; before then, I had only known arithmetic.

By the time I entered high school, my father's health had become unstable; it was unclear whether the cause was overwork or if it was residual of a case of influenza he caught during the 1917 epidemic. On one occasion, his doctor sent him off for treatment at a rest home. He returned looking much healthier, but his recovery was not permanent. In 1923, when I was a sophomore, he was diagnosed with tuberculosis, so he resigned his pastorate and went to a sanitarium in the Adirondack Mountains. From his bed he wrote me several thoughtful letters about my growing up. My mother and I went up to visit him and he seemed to be cheerful and improving, but I could not have realized that I would not see him again.

Chapter Two
High School

When my father became ill, my grandfather, William Ward McLane, still held his pastorate in North Leominster. His wife Frances, who suffered from what may have been Alzheimer's disease, had recently died, leaving him alone with a maid in a large parsonage. He offered to take in my mother and her three boys, and she gladly accepted since there was no disability pay or pension to replace my father's lost salary. I helped to pack up our possessions, which included many books; among them I remember specifically the 11th edition of the Encyclopaedia Britannica, with its splendidly accurate and comprehensive information—I still use it to this day.

After we settled with my grandfather, my mother found a job teaching English at North Leominster High-school, and raised Gerald and David as well as she could under the gaze of a father-in-law not attuned to the needs and ideas of little boys. I began to feel a little more independent, and admired the way my mother managed her difficulties. I joined a troop of Boy Scouts, attended a group for young men in Sunday school at my grandfather's church, and entered school as a sophomore.

North Leominster High School was an exciting place for me. As a sophomore, I learned Euclidean geometry. However, the school evidently considered this subject a bit too difficult for most students— as juniors we all reviewed and repeated the course in geometry. Our teacher, a young woman, chose to send the students to the blackboard

Sunday school class, Leominster, MA: Saunders is in the third row, second from the right

to present the theorems of the day. I recall one occasion involving a theorem about a triangle: I knew that the specific shape of the triangle did not matter, and that its vertices could be lettered at will. So when I went to the board, I drew the triangle upside down, changed the letters labeling the vertices, and presented the proof at flank speed, all to the evident distress of my teacher, as some of the students (my friends) egged me on. In retrospect, it is apparent that I understood the proof from my first geometry class, but that I did not at all see how to communicate the proof to my fellow students; I must have been a real nuisance to the young teacher.

At that time, students who had study periods were assigned to sit in the back of active classrooms. After one of my impetuous performances, an older student with a study period in my class came by my desk to leave a brief note that said, "Keep it up," signed, "A math tutor." The note was written by Earl D. Rainville, who was eventually a mathematician on the faculty of the University of Michigan, where he prepared textbooks on differential equations and special functions. In subsequent years, we were acquainted only distantly, and I regret this loss of contact.

The Boy Scout troop provided good fun with hikes and cookouts, and there was also an annual competition between the troops in the neighborhood—this I enjoyed. There was one competition about

Chapter Two — High School

Saunders with his brothers
Gerald and David, Norwalk, CT, 1927

knots, in which each contestant carried four short ropes and ran across the room to a log, tied on the ropes using different knots in a specified order, and ended with a bowline that he used to pull the logs across the room. I can't recall that I ever won, but I did learn those knots. However, I did not yet know of the many mathematical questions associated with knots.

We also competed in Morse code signaling: The signalers stood on one side of the room, sending the prescribed message by the usual successive position of the flags. My signaler was athletic—he signaled fast. I sat on the opposite side and quickly read off the letters to my scribe. We won often. As I recall, we even floated a rumor that we had set a world record. Of course, we never really pressed this issue; it served mostly to bolster our self-esteem.

In later years, I returned to Leominster for a class reunion. I found that my signaling companion ran a summer camp for youngsters, and so had realized his remarkable athletic capabilities; however, we did not keep in touch further. I did manage to keep up with other friends: John and Bill Grubb; Wallace Gove; and Ralph Kirkpatrick, who became a musicologist and a famous performer on the harpsichord. He was most famous in the 1950s and 1960s for his reinterpretations of Dominico Scarlatti. Ralph and his sister, Annis, were great friends with my mother; Ralph kept a harpsichord at her house for practice.

During my high-school years, the Ku Klux Klan was active and apparently critical of Catholic doctrines. A Congregational church downtown received and praised Klan members with these views, and the excessive attacks on Catholicism displeased my grandfather. He proceeded to open his church on a weekday to give a two-hour lecture on the origins of the Protestant/Catholic divide, which was well-attended and impressively done. I admired his devotion and vigor in defense of tolerance, and marveled at his decisive activity at nearly 80 years of age.

This does not mean, however, that I understood his ideas. I did join his church, and in the process of doing so, remember being questioned carefully by an elderly member about my beliefs. I answered seriously, but at the same time, kept some reservations about certain points of doctrine. I struggled with aspects of my ministerial heritage, but I did not often approach my grandfather for advice and wisdom. On one occasion, I asked him about the purpose of individual life. He responded that we were there to exhibit the glory of God; this conclusion stopped me cold—God's glory was not visible to me.

At the high school, my English teacher was Olive Greensfelder, a native of Hyde Park in Chicago, who had come east to do graduate study in education. She took much care in reading and criticizing my writing; I specifically remember an enthusiastic essay I wrote about Abraham Lincoln. I consider her guidance in writing to have helped me later in the exposition of mathematics for students. She established a high-school newspaper and named me the editor. The first assistant editor was a beautiful blond named Helen Wolcott, whom I admired, but only at a distance. I did learn to dance, and had occasional dates, but not with Helen, as she was dating my best friend, Merrill Bush.

The high-school boys were organized into companies for required military drills, which we carried out on the football field in uniforms that echoed Civil War style, caps and all. The drilling, I suppose, was

Chapter Two — High School

good for our discipline. We learned the manual of arms, and I even rose to be second lieutenant of my company. Each spring, at the town exercises for Memorial Day, the four student companies paraded, and there was an annual competition for "best company."

My father's pacifism during World War I was still on my mind: One year, on the day of the annual drill, the high-school newspaper carried my editorial criticizing the whole military drill requirement. It caused considerable discussion. The faculty member in charge of the drill, however, responded only mildly, pointing out that I had split an infinitive. The drill continued despite my editorial, but I imagine I felt that I had supported my father's views.

Early in 1924, we received the devastating news of my father's death at the hospital in the Adirondacks. At first, we did not tell my brothers. David was only 2 or 3, and Gerald was about 7. As I mentioned, my father had kept in careful touch during his hospitalization, writing me thoughtful letters about the challenges of growing up. I tried, in small ways, to express his influence to my brothers, but did so imperfectly. Gerald finally asked me when father would come home, and at the age of 16, I did not know how to explain such things to him. I only said, "He's not coming home. He's dead." Learning the truth in that way from me was very traumatic for him. I did realize that his death was a terrible blow for my mother, but I was not always adequately helpful. For example, one Halloween I rang too many doorbells and a policeman called at the parsonage to explain my misbehavior to my grandfather. He must have found these times troublesome, but he managed to accept me.

I enjoyed relationships with my father's full brothers: my Uncle Stanley was an engineer who worked at General Electric, and my Uncle William was an enthusiastic salesman. I had yet to develop relationships with my father's half brothers, Uncle John and Uncle Paul, who were sons of my grandfather's first marriage. But during a visit, in 1924, Uncle John turned to me and said, "Saunders, I will send you to Yale." Yale was a natural choice since our family had a history of attending school there, including all five uncles and both of my grandfathers, as well as earlier relatives, such as a distant cousin

Early Years

Ami Ruhami Robbins. I eagerly accepted Uncle John's offer, of course. As will soon appear, I profited much from my undergraduate years at Yale.

At this time, Uncle John was a senior partner of the Wall Street firm Simpson, Thatcher, and Bartlett; he was married, but with only one child, an adopted daughter. His generosity in providing higher education for me, for my brothers and for others as well was remarkable. Most of my college-bound high-school classmates were headed for Harvard, but I was happy to apply to Yale. I assume that my English teacher, Miss Greensfelder, wrote well about me, but presumably I would have been admitted anyhow as a "legacy," an idea then and now popular in the Ivy League.

Uncle John also observed, and correctly so, that I was a naive and inexperienced young man. After my junior year, he arranged surveying work for me during the summer with one of his clients, the Pennsylvania Power and Light Company. The company had recently established a dam for water power, creating Lake Wallenpaupack, and they needed to survey its boundaries. I was a rodman in the survey gang. Each morning we drove out to the site of the new lake and proceeded to survey more of the resulting property lines. I dressed in high boots for protection against rattlesnakes, and carried a brush hook—an ax with a hooked blade. I cleared away all of the brush in the line of sight so that the surveyor could see the next stake from his transit. I learned to whack brush and to hold a pole steady for sighting on the true vertical over a stake. I was fascinated by the way the surveyors used trigonometry, which I had not yet studied thoroughly at school. The transit, which is a telescope mounted to swivel both horizontally and vertically on a tripod, is a wonderful instrument for measuring angles.

However, I was alone in the company of older men and their raucous view of life. When the day was over, we congregated in the bar of the local hotel, in the town of Hawley. I had been trained that drink was evil, so I found myself in an uncomfortable new experience. My friends—the survey gang—kidded me mercilessly. I was 16 and homesick, and I almost gave up. But a wiser, older man listened to my troubles and made some useful adjustments. I was moved from

Chapter Two — High School

the rowdy hotel to a room in a private home where I had more freedom—I was able to do things like play records. I seem to recall playing the record "Forsaken" over and over again. That was how I viewed myself. Somehow, I stuck it out for the summer, finally returning to my grandfather's house. I constructed a partly working model of a transit so that I could explain to my brothers what I had been doing. I do not recall trying to explain my homesickness; this was my private trouble.

The following summer, I did not return to surveying in Pennsylvania. Instead, I got a job in a local celluloid factory, where I spent most of my time beveling the tongues on barrettes for ladies. Since I stayed at home, there was no problem of homesickness, and I did earn a bit of money; I also learned how dull and repetitive work in a factory could be—a useful bit of information, to be sure.

Chapter Three
Undergraduate at Yale

In the fall of 1926, at 17, I went off to Yale. My mother arranged for me to room with Harvey Morrison, whose family we knew. I was initially startled by some aspects of student life, which was dominated by students who had come from famous prep schools such as Exeter, Andover, and Kent. I quickly learned that a proper Yale man should be active; he went out for football, managed an athletic team, or "heeled" the *Yale Daily News*. I did not even try to go out for the *Daily News*; I concluded that my experience with a minor high-school newspaper would be of little help. I did once try out for track with no success—I was not fast enough. In general, I avoided the sanctioned activities for students and concentrated on the excitement of the ideas that came up in my courses.

Shortly after arriving, a representative from the well-known New Haven department store J. Press called on us in our rooms to offer a "Pressing Contract." I did not think I had enough money for this, and regardless, the idea of freshly pressed pants for each day did not seem necessary to me. I did, however, follow the common practice of packing up my dirty laundry every week and mailing it home to my mother to be washed. As for money, my Uncle John provided me with $1,200 each of my four college years to cover my tuition and other expenses. This amount sufficed, but I am struck by how ridiculously small it seems today.

Before coming to college I had considered my financial future—it was clear to me that my family had experienced difficult financial times. I hoped for some kind of intellectual career, and carefully calculated what one needed to save for old age in those times: $100,000 at four percent interest would yield $4,000 per annum, which seemed to be enough. But what career would be possible? In high school I enjoyed physics and chemistry, even though the teacher was not very expert (he could not explain the gas law $PV=kT$ or the meaning of k). I found a book by S. Slosson that described the developing uses of chemistry in industry. This, it seemed to me, offered a career in science that might provide for my desired savings of $100,000. I was not yet aware that there was such a thing as a career in mathematics (teachers in my high school did not appeal to me as role models).

For these reasons, I intended to major in chemistry. As a freshman, I was put in an advanced section of chemistry that actually dealt with qualitative chemical analysis. But I found the laboratory a bit dull—beakers and other such equipment did not much appeal. I also took calculus, where I discovered the remarkable world of derivation and limits, as well as Newton's wonderful use of calculus to describe the planetary orbits. My teacher, an instructor named Lester Hill, was a graduate student working toward his Ph.D.—today he would be called a teaching assistant. He encouraged my interest and talent in mathematics. In the spring I took the Barge Prize exam in mathematics, and won. I immediately invested the $50 prize money in the purchase of a pocket watch. In those days, a proper Yale man wore a vest with a watch and chain; Yale students were, of course, trained to be very proper.

At this point, I inquired about a career in mathematics and learned that one could take exams and become an actuary with a job in an insurance company. This news sufficiently fitted my financial plans, enabling me to switch from chemistry to math as a major. I owe this decision chiefly to Lester Hill and his exciting introduction to calculus. It seemed to me then (and does now as well) that calculus involved deep ideas. Our text was *Calculus* by Longley and Wilson; it was somewhat similar to the later popular calculus text by Granville (eventually, Granville, Smith, and Longley). During that period, the Yale faculty held that basic courses should use texts written by Yale

Chapter Three ~ Undergraduate at Yale

faculty members. William Raymond Longley held a professorship in the Sheffield Scientific School (largely meant for engineers), and Wallace Wilson was a professor of mathematics in Yale College—I later knew him through other connections.

My other courses included English and German, the latter a beginner's course I took because of the importance of German books in the sciences. My instructor, Professor Schreiber, was interesting, and I found the German language attractive. I recall little or nothing of my course in English.

In keeping with Yale traditions, I made a number of close friends. Donald Wright, a New Haven native and chemistry enthusiast, became my roommate sophomore year. Bob McKenzie and Pearson Hunt came from one of the minor prep schools. Pearson was a geography enthusiast; we purchased an atlas and studied the countries of the world. During our sophomore year, Bob had a breakdown and left college, which left me mystified and troubled. I still keep in touch with Pearson, who became a professor at Harvard Business School.

At that time, Yale students were all registered in a common freshman year, after which they were to choose between YC (Yale College), or Sheff (the Sheffield Scientific School). I chose the College, where I knew that I could major in mathematics. Because I had dropped Latin after only two years in high school, I could not be a candidate for the B.A. degree. Instead, there was a Ph.B. degree, which required a year of study of classical civilization in a course designed to replace some of the vital knowledge of Greek or Latin; it was a good course, with a talented teacher.

In high school, Ralph Waldo Emerson had fascinated me. During my sophomore year at Yale I took a historical course in philosophy from a pompous professor, who repeatedly told us of the importance of "old Paremenidies." I found this dull. But soon I came to know Filmer S. C. Northrop, a young professor of philosophy and logic. He had just arrived at Yale after studying at Harvard with Alfred North Whitehead; he was an enthusiast for logic, claiming that if one understood the premise of any book, one could logically predict the conclusion to be drawn—an exciting idea, though it now seems to me a bit too simple. I audited a course that Northrop taught and talked a good bit with his student, John C. Cooley. In this vein, I learned about the then new book *Principia Mathematica* by Whitehead and Russell. In three volumes, it proposed a firm foundation for all of mathematics on the principles of logic.

How wonderful! I bought and annotated a copy of the first volume. Like many other readers, I never did get around to buying or reading the second and third volumes, but the first made a major impression. All of mathematics founded on logic—wow! I did not at first realize the subtleties of Russell's ramified type theory, and I certainly did not anticipate category theory and its relation to set theory, but I became fascinated with it and with the idea that there could be strict and dependable rules of proof.

Sophomore year I learned a bit more calculus from James Pierpont—the *éminence grise* of the mathematics department. I recall that he was bored teaching calculus to sophomores, often complaining about his salary and his perquisites, and the ragged lining of his coat, which he displayed. In his prime, Pierpont had been an influential mathematician, writing texts on complex and real analysis and giving colloquium lectures on Galois theory to the American Mathematical Society.

I had the opportunity to spend the next year abroad in Germany under a well-established program, Junior Year Abroad. With regret, I turned down this opportunity. My professor of classical civilization criticized my decision; he thought I was much too wedded to the narrow field of mathematics, which did not seem to me all that narrow.

I also took a course in physics, skipping the freshman physics to take the sophomore level—it was great. The principle teacher was Jesse W. Beams, who was even then building the high-speed top that later made him famous. He guided his students with enthusiasm more than with precision. For the laboratory work, my partner was Julian D. Ripley, a socially well-placed, enthusiastic student. Our lab instructor was a smiling young man who urged us to do something a little out of the ordinary—why not measure the value of the constant gamma for air? We had fun measuring it, but I do not recall that we

Chapter Three ~ Undergraduate at Yale

got the correct value, then thought to be 1.402 ±0.001. I guess we pleased our lab instructor, Ernest O. Lawrence. He left Yale after that year for Berkeley, where he built his famous cyclotron. Ripley and I had a splendid opportunity to learn from a master about laboratory work in physics; sadly, neither one of us became professional experimentalists, though in later years, Ripley wrote a popular text on physics for undergraduates. Long after our class with Lawrence, I interviewed as a candidate for the presidency of Yale (Uncle John suggested my candidacy). Most of my interview involved questions about Lawrence, who was also a candidate for the position.

My lab partner Ripley had a wide-ranging desire for all types of learning, and developed an appetite for universal knowledge—the organized understanding of all that is known. As a junior, he had a double major in philosophy and physics. He told me of the exciting physics honors course given by a European physicist, Kovarik, an expert on radioactivity. It was indeed an exciting course, so I signed myself up for a double major in math and physics, and enjoyed Kovarik's course. I was also much taken by the idea of a possible universal knowledge. I believed that a proper academic ought to know everything, but of course, I never did succeed in mastering universality, nor do I know exactly what this might have meant.

Yale College offered a choice of courses: I also took Economics 10 as a sophomore. The systematic text by Fairchild, Furness, and Buck (again, all Yale professors) was careful to include a chapter on the economic function of government, listing those few things the government could properly do without harming individual initiatives. In other words, the text clearly had a conservative slant. The text also considered marginal cost; I recall a long and pedantic explanation—that really amounted to a description of marginal cost as a derivative—given for students who had not studied calculus. I found the explanation to be too long-winded.

This was just before the stock-market crash in 1929. I was not yet aware of Keynesian economics, but I was curious about socialism, so I audited Professor Furness's socialism course. It was an objective and carefully historical presentation, but clearly designed not to encourage fellow travelers. All told, I was able to learn much about economics, its influence, and its use of mathematics, but it was not quite enough to shift my main interests from mathematics to economics. In my junior year, I started to take a course in accounting. I thought this would please my Uncle John, who might have been inclined to send me to law school. However, the accounting course turned out to be dull and routine. So I was not converted to the business life or to the ideas of socialism; the ideas of mathematics, though more austere, were, for me, more powerful.

At the start of my junior year I approached Professor Wallace Wilson, a senior member of the mathematics faculty. I hoped that he would support a reading course for me, perhaps in Whitehead and Russell's *Principia Mathematica*. He was disinclined to accept my choice; he probably shared the then-general opinion of American mathematicians that logic was not really a part of mathematics (how could logic be the foundation when mathematics preceded Whitehead and Russell?).

Professor Wilson proposed instead that I study set theory from the famous book *Mengenlehre* by Felix Hausdorff. I did study it, and profited much, learning about cardinal and ordinal numbers as well as set theory and metric spaces (a specialty of Wilson's). *Mengenlehre* was my first real mathematics book, that is, one that was not a textbook. It proved to be a very good introduction to real mathematics from the hands of a real master. Inevitably, I developed a good intuitive feeling for Hausdorff spaces, a subject that soon became basic to all manner of mathematics. It is interesting to speculate on whether Hausdorff influenced more mathematicians than Whitehead and Russell.

That year, I also came to understand mathematical precision and rigor in the course on advanced calculus taught to juniors by Professor E. J. Miles. We used an authoritative textbook by Edwin B. Wilson (I later knew him well in connection with the National Academy of Sciences, where he was the managing editor of the Academy proceedings). Wilson's textbook presented a rigorous view of the basic ideas of the calculus. He did not overemphasize—at one point he advocated "vigor" over "rigor"—but he gave a clear presentation of all the epsilons and deltas needed for the standard descriptions of

Chapter Three ~ Undergraduate at Yale

limits. I had probably heard of them before, but his book's presentation firmly fixed them in my mind. They may not have been so well-fixed for my fellow students: Professor Miles actually exempted me from attending the sessions of his course. Instead, I graded the other students' papers, surely an odd arrangement that the students might have protested; perhaps they did not know I was grading their papers. In any case, student protests were unheard of at that time.

Professor Miles was a dynamic, hands-on teacher: he wanted all his students to learn, and was not above pounding a student's shoulders to encourage his learning. I did not adopt this technique, but I learned many other things from him about effective teaching, such as the importance of paying attention to individual students. He had written a beginning calculus textbook with E. J. Mikesh, and hoped to supplement this with a good text of advanced calculus, since at the time there were few available in English. After I graduated in 1930 he enlisted me in the project, so I spent the summer of 1930 in New Haven, composing pieces of a text with him. It was not an appropriate task for me then; I had not taken any standard course in real variables, which really develops advanced calculus. I do recall desperately trying to gain that knowledge from reading Goursat's famous *Course Analysis*. During that summer, Professor Miles and I wrote many chapters of a possible book, though it never was completed or published (I left for graduate work in Chicago). But I did learn a good deal: Professor Miles had a remarkable talent for exposition, and I learned much that I would use later in my own writing. I regret that my help did not bring his proposed text on advanced calculus to a real conclusion.

In the summers after my freshman and sophomore years, I returned to work in the survey group of the Pennsylvania Power and Light Company—by this time, homesickness was no longer a problem for me. But after my junior year, I took other summer employment at the General Electric Company in Schenectady, New York. The famous chemist Irving Langmuir was in the G.E. Laboratory at that time; one of his nephews, a classmate of mine, proposed that I work there for the summer. I had a positively experimental job measuring the lead losses in tungsten filaments in light bulbs. I did not manage the apparatus very well, so my results had to be redone in a joint paper with Langmuir and Blodgett, published in the *Physical Reviews*; it was my first publication. Again, this experience confirmed that experimental physics was not really my forte. However, the summer gave me a fine opportunity to watch Langmuir's scientific curiosity spin at full throttle, as for example, in measuring the currents in the lake near his summer cottage. This job was much more fun than cutting brush for the Pennsylvania Power and Light Company, and it gave me a broader perspective on science.

While I was in college, my family situation had changed. Marjorie Harrington, my mother's cousin, lived in Norwalk, Connecticut. Her parents had died, leaving her a fine house built by her father. She offered to take in my mother and my two brothers, and cheerfully extended the invitation to include my mother's mother, Isabel Saunders, who was then in poor health. Her proposal came at a fortunate time, as my grandfather was about to retire from his church in North Leominster and would have to move out of the parsonage. So my mother, Gerald, and David settled in with Marjorie. My mother was soon teaching mathematics and English at a number of nearby private schools for girls. Marjorie, who loved children, took a lively interest in my brothers. Norwalk was close to New Haven, so I looked in from time to time; I was not always comfortable with the way the household was run, but the essential aspect of the situation is that it relieved me of a potential obligation to support my family. As a result, I was able to plan to pursue graduate work in mathematics.

My grandfather retired and lived in Lynn, Massachusetts, near his son Stanley. I never had the courage to tell him how much I admired his independence from popular views.

I also made a few visits to Leominster to see friends and to date one of my fellow students who worked with me on the school newspaper. Alas, dating at a distance did not turn out to be very convenient, and the young woman in question was soon writing me about her current boyfriend. I once visited the Connecticut College for Women, and later, Smith College. The latter trip was for a meeting of the Model League of Nations. One young woman there accepted my invitation

Chapter Three Undergraduate at Yale

Attains highest accademic standing in the history of Yale: Saunders Mac Lane of Norwalk, who maintained an Average of 384.8 in his studies for the first two and a half years at the University, for which he was honored by being elected President of the New Haven Chapter of Phi Beta Kappa to the Yale Prom. She later invited me to visit her home in New York, where she explained how some of her friends had been financially devastated by the 1929 crash of the stock market, a crash I had not even noticed!

That lack of attention to the outside world was a natural aspect of my life as a senior at Yale; the search for knowledge and friendship left no room for observing the stock market. Donald Wright and I had a large suite on the seventh floor of the splendid new Harkness Memorial Quadrangle (HMQ). In the morning we would rush down those six flights of stairs to get to class; in the evening we watched the main entrance of HMQ to see the acolytes of the famous secret society, Skull and Bones, returning in lock step from their secret meetings.

Emphasis on tradition was still powerful at Yale, evidenced by the prevalence of the senior societies such as Skull and Bones and Scroll and Key. The famous Tap Day came at the end of junior year. All of the juniors congregated in the old campus hoping to be selected while the senior members of the secret societies walked among the juniors tapping selected candidates and saying "Go to your room." I stood there with all my classmates. As a junior Phi Beta Kappa I must have considered that I was worth a tap, but it never arrived. I wept no tears—I had aimed my undergraduate career in the appropriate direction. I cared more for lessons than for student customs. As a sophomore, I declined to join a fraternity; as a senior, I would not have made a loyal member of a secret society. Make no bones about it, any education involves choices, and a college student must choose.

Chapter Four
Universal Knowledge and New Knowledge

At the start of my active interest in a mathematics career, I do not believe I was aware of research that lead to new results. Mathematics was fascinating, and much of it new to me. Calculus, for example, was exciting, but it seemed as though it had long since been entirely worked out. I remember well my course on the beautiful subject of theoretical mechanics with E. W. Brown, an expert on the motions of the moon. He lectured to us from dog-eared notes on classical mechanics and the deep ideas of Hamiltonian mechanics. This subject, apparently, had also been all worked out. My brief acquaintance with the work of Whitehead and Russell suggested that there had been a recent period of exciting and new results in the foundation of mathematics and logic, but those massive volumes convinced me that this area had been fully developed as well. It was all there: Professor James Pierpont's course in sophomore mathematics gave little trace of original study; advanced calculus was good and deep, but the problems were just exposition of known ideas; and Cantor's work on set theory and Hausdorff's on topological spaces had been new, but seemed largely finished.

Such was the tenor and style of teaching in all Yale undergraduate courses. When we listened to the polished lectures of William Lyon Phelps on Tennyson and Browning, one of my classmates used his father's notes from an earlier version and found that they were almost identical. I dismissed Phelps as superficial—he lectured well but had no new ideas. Instead, I audited a course on the Age of Johnson with the eminent Chauncy Brewster Tinker. He was more astute than Phelps, a real scholar, but still not an innovator. We did learn about the use of calculus in economics, which was something new, and sciences such as chemistry and physics involved current research, but it all seemed experimental and despite the efforts of E. O. Lawrence, I was not inclined to do experimental work. Surprisingly, I did not learn about the new results in quantum theory as an undergraduate.

It seemed as if all of our attention was directed toward knowledge that was already known; therefore, during the first years of my undergraduate education, I put my own emphasis on acquiring universal knowledge—the assimilation and organization of everything known. Then, something exciting happened. In the fall of 1929, my senior year, the Yale Mathematics Department had a new, young member—Assistant Professor Oystein Ore. Ore was a Norwegian mathematician who recently studied in Emmy Noether's school of abstract algebra in Göttingen, Germany. At first, I did not notice his presence, but my fellow student and close friend Eugene Northrop did, and soon told me that Ore was teaching two graduate courses: one in Galois theory and one in group theory. I came to listen, and discovered the developing ideas of modern abstract algebra. Ore's presentation was perhaps a bit hasty, but he succeeded in conveying the excitement about the structure of groups and Galois theory.

A group was a collection of things, any two of which could be multiplied: Galois theory showed that the n solutions of a polynomial equation of degree n formed a group. The abstract properties of this group were then used to determine which equations could be solved by explicit formulas, like those for a quadratic equation. Galois theory was profound, but at that time the standard American texts treating it were generally clumsy and confusing, as can still be verified. With Professor Ore's guidance, I promptly bought and studied several of the current German texts in algebra. Thus, my earlier study of German paid off. One of the books I studied was the classical multivolume text by Robert Fricke. There was another, more adventurous text on modern algebra by Otto Haupt, a professor at Erlangen, a university near Nürnberg, now famous for Felix Klein's "Erlanger Programm." Haupt was well acquainted with the novel ideas of Emmy Noether, who had come from Erlangen where her father, the famous Max Noether, taught. Haupt's two-volume text was then at the extreme edge of abstraction, an abstraction I enjoyed much even while I found it excessive in some points. I learned a great deal from it. I believe, however, that Haupt had little general influence; the breakthrough of the modern approach came with the publication of the much more elegant *Moderne Algebra* by Van der Waerden, which was not published until 1931.

Ore's lectures and Haupt's book intrigued me. The observation that abstract group theory could be used to analyze the situation of polynomial equations was amazing. Here was something in the way of new knowledge developing in mathematics. I quickly came to imagine that there were many other algebraic situations that could be neatly axiomatized and clearly understood (I actually did try that in my master's thesis the next year at Chicago). But it was not just this one example of axiomatic clarification that I gained from the new approach: the work of Emmy Noether and her successors indicated to me that there were brand-new ideas to be found in mathematics. With this indication, my focus shifted from the accumulation of knowledge to the hope of discovering new knowledge. To be sure, this was happening in many parts of science, but in mathematics I could see that there were many more new things to be done—I had begun to understand what a research university could be.

In retrospect, it is strange that I made this discovery in mathematics rather than physics. I continued my double major in mathematics and physics; this meant taking Professor Leigh Paige's systematic course in theoretical physics, which met five days a week and was designed to give beginning graduate students a comprehensive start. Professor Paige had just reduced his course to a textbook, so his lectures closely followed the book—they of course began with an exposition of vector analysis in the spirit of Gibbs-Wilson, whose decisive book on vector analysis had been written at Yale. The preface to Paige's book observed "the recent atomic theories of Heisenberg and Dirac and Schrödinger have been omitted as being placed outside the scope of this book by their mathematical complexities." This was 1929, when modern quantum mechanics was just entering the scene, but Paige chose not to include it. Because of this omission, my own discovery of the central role of research came not in physics, but in mathematics. It was not until much later that I really understood the general importance of research in the development and organization of knowledge in all fields. I now believe that effective teaching that motivates and stimulates students into creativity grows out of research and the incorporation of new ideas into the curriculum.

My new realization of the importance of research coincided with the ongoing changes at Yale. As a freshman, my original impression reflected the university's role as a training ground for privileged young men through classes and student activities. An extreme interpretation is superficially expressed as follows: one made student friends at Yale and then went to Wall Street to sell bonds to these friends. This tradition was amply present, but there was also a new emphasis on scholastic accomplishments. James Rowland Angell had recently been appointed as president; he was not a Yale man, rather, he came from the University of Michigan. In the evenings I often observed him walking home from his office in Woodbridge Hall, deep in thought. I could not have known his thoughts, but it seems now that President Angell understood the nature and the future of what would now be called research universities. I cannot document the steps he took to emphasize research at Yale, but I am convinced that the presence of new, young faculty served to help redirect the university, and my own profitable contacts with new faculty are clear examples of this. At the time, I could not have seen that this was the beginning of a transformation that would make Yale a research university.

Many years after I took the course on theoretical mechanics from E. W. Brown, I chanced to meet him at a mathematics meeting. Before I knew him, he had spent many years calculating, by hand, the future motions of the moon. When I met him again he told me that, using modern computers, all of these tables had been recalculated, revealing few corrections. He did not seem troubled by the unanticipated speed with which his life's work was replicated through new technology: perhaps he realized that new techniques growing out of research and development would help to confirm and expand our knowledge.

Chapter Five
The University of Chicago, 1930

At the end of the Yale football season every fall, the Yale Club of Montclair, New Jersey, held a party to celebrate the achievements of the Yale football team. By the summer of 1929 (before the stock-market crash), general prosperity suggested a wider celebration, so the club offered a new award: in addition to honoring achievements in football, the club would present a cup to the man who had "made his Y in life." In 1929, Robert Maynard Hutchins accepted the Presidency at the University of Chicago after serving as Dean of the Yale Law School; thus, he had clearly "made his Y in life." However, not all students in New Haven were in football, and so there was to be another cup for the student who "made his Y in scholarship." My grades had been high, with an overall average asserted to be the "highest ever" (though "ever" represents a long time), and I was chosen by Montclair for the award—I still have the cup.

Receiving this award meant that I met R. M. Hutchins at the Montclair Yale Club party in 1929. He asked me what I planned to do after graduation, and I responded that I intended to pursue graduate studies in mathematics. "Come to Chicago," he said. "We have a great mathematics department." Some weeks later, a letter from Hutchins offered me a fellowship with a stipend of $1,000. My teacher, Egbert Miles, had done his graduate work at Chicago, so he was pleased with the offer. I accepted.

My choice did not please Oystein Ore. He was evidently knowledgeable about the state of graduate study in mathematics at

Early Years

American universities, and was happy to point out to me that Harvard or Princeton would have been better choices (I suppose because of faculty members there). It may be that Ore had hoped I would stay at Yale, but the graduate students I knew at Yale did not especially impress me then. In retrospect, I was wrong. One of those students, Grace Murray Hopper, later worked for the Navy (eventually becoming an admiral) and was a pioneer in the development of computers. The story is told that Grace coined the term "bug" to refer to an unexplained computer problem after a moth flew into the circuits of an early computer.

In September 1930, I reported to the Mathematics Department at Chicago to begin my fellowship. It turned out that although President Hutchins invited me, I was not formally admitted to the graduate school, and the department did not know about my fellowship. Professor Gilbert A. Bliss, the chairman of Mathematics, was annoyed that he had not been consulted about me. He examined my transcript and was promptly dubious about giving me credit for a Yale course with James Pierpont with the vague title Sophomore Mathematics. Nevertheless, I was admitted to the graduate school and enrolled in suitable courses at Chicago. My troubles were not terribly surprising; Hutchins was young and had been appointed president quickly, which did not sit comfortably with many older Chicago faculty members. But I did learn from that faculty, as follows.

Eliakim Hastings Moore was the first head of the Mathematics Department of Chicago, in 1892. He trained many famous mathematicians, including Oswald Veblen (Princeton), George D. Birkhoff (Harvard), and T. H. Hildebrandt (Michigan). He was no longer head when I arrived, but he continued to be active in the department. I joined his seminar, which intended to deal with the Hellinger integral, something I never did understand. I told him about my interests in the foundation of mathematics, and he at once suggested that I report on a paper by E. Zermelo, giving axioms for sets. Zermelo-Fraenkel axioms were, at that time, the chief alternative to the foundations of mathematics proposed by Whitehead and Russell. I reported on the paper with enthusiasm, especially in explaining how Zermelo used the axiom of choice to prove that every set could be well-ordered. I thought I had done very well, but as soon as the (few) others had left the room, Professor Moore took me aside and explained what the paper was really all about and what I should have said. That was an occasion in which I learned a great deal thanks to Professor Moore: I learned from him how to give a talk on mathematics, and I learned about sets as a foundation for mathematics. He was an amazing professor.

Leonard Dickson, the famous Chicago algebraist, was giving a course on one of his special interests, number theory (he had written a masterful *History of the Theory of Numbers*). He had us read his new expository test, line by line, and insisted that we pay careful attention to his choice of notation, and of course, we did—he was an impressive authority. On one occasion when he tipped over backwards in his chair we did not dare to laugh. His research at the time was a detailed study of some aspects of the Waring problem. Every integer can be represented as the sum of at most nine cubes. In general, the Waring problem is to determine, for a given positive integer k, the smallest number of terms required to represent every integer n as a sum of kth-power terms. His detailed calculations supplemented the asymptotic results on higher powers then being studied by analysts such as Edmund Landau (from Göttingen). Landau came to visit and lecture; at the start of his lecture, he observed that there was no one equipped with water and a sponge to wipe off the blackboards as in Göttingen. Professor Bliss immediately assigned this task to me. My new duties did not seriously impede my limited understanding of Laundau's lecture, which, according to his habits, was very precise. I decided that Landau's analytic results about all sufficiently large numbers were more exciting than Dickson's calculations for numbers not so large.

Bliss himself was perhaps the world's leading expert on the calculus of variations, a topic that had been firmly established in Chicago by Professor Oscar Bolza, a member of the original Chicago department. By 1930, the subject was a chief source of Ph.D. theses at Chicago. I took the course that Bliss gave for students writing theses on this topic, and I was somewhat perplexed to see that Bliss didn't much insist on careful use of epsilons and deltas (important for the particular "sufficiency" theorems of the subject). However, at one point I asked him for details and was much impressed with his mastery of these matters; he evidently did not wish to distract all of his students with issues of detail.

Bliss also liked to tease his students: He emphasized his own first mathematical visit to Paris, and how the students all stood up to welcome Professor Hademard to the podium. Chicago students, Bliss said, would never do that. My fellow students disagreed. It so happened that I was the only student who had a tuxedo, so the next day I wore it to summon Professor Bliss to meet our class, and they all stood up to welcome him—he enjoyed it.

Bliss continued to guide many graduate students (with or without the use of epsilons) to work in the calculus of variations, but that subject was changing from its traditional form. At Harvard, Marston Morse—a student of George Birkhoff, and hence a "grandstudent" of E. H. Moore—had developed the "Morse relations" between connectivity and extreme points. For example, consider an imaginary chain of Hawaiian islands: count i the number of islands; l the number of lakes; and p_0, p_1, and p_2 the respective numbers of peaks, passes, and pits in the whole island chain. These numbers are related; for example, each island must have at least one mountaintop, so p_0 is greater than or equal to i. Other relations can be seen by flooding the whole island chain and then counting what happens as the flood gradually recedes. These and other relations between these numbers are the Morse relations. They connect the numbers of maximum, minimum, and saddle points as studied in the calculus of variations with the so-called Betti numbers (the number of islands, lakes, etc.) that come from algebraic topology. Other experts on the calculus of variations had not expected any such connection. Moreover, the Betti numbers were not just numbers, but were actually invariants of certain abelian groups.

Morse applied these relations extensively in his most recent book, *The Calculus of Variations in the Large*. Professor Bliss was eager to understand how these new ideas could influence his subject of the calculus of variations, so he assigned me to talk about topology (the

Chapter Five — University of Chicago, 1930

Betti numbers) in his seminar. I knew virtually nothing about topology then, so I studied the most accessible source, *Analysis Situs* by Oswald Veblen. This was my first contact with Betti numbers, but Veblen did not trouble to explain that they were invariants of certain homology groups of spaces, and I did not even know about the crucial Morse relations.

I learned other parts of advanced mathematics at Chicago, such as projective differential geometry from Professor E. P. Lane, and many aspects of theoretical physics from Professor Lunn. Professor Lunn had a wide knowledge of applied mathematics, but his presentation was somehow hampered by his recollection that his 1928 paper in which he independently discovered quantum mechanics had been rejected for publication (I do not know what might have been in that paper). I also learned a great deal of algebra from Professor R. W. Barnard, a protégé of E. H. Moore. For Barnard, and for Chicago in general, a vector was an n-tuple $(y_1,\ldots,y_n)$ of numbers, an idea that I soon had to unlearn.

Professors Moore and Barnard were well practiced in Peano's logistic notation: $\wedge$ for "and"; $\vee$ for "or"; $\exists x$ for "there exists an x"; and $\forall x$ for "for all x" (the quantifiers). All theorems were formally stated in Peano notation, but their proofs were casual and informal. Whitehead and Russell used Peano notation for statements in *Principia Mathematica*, but had added formality for proofs. I was troubled by Barnard and Moore's deviation from the formal idea of proof I had learned from Whitehead and Russell. At that time, I felt that proofs had to be precise: I wished to combine the notion of a proof as a precise expression with the understanding of the idea of that proof. I recalled the elegance of German abstract algebra, where rings and fields axiomatized the properties of addition and multiplication. My master's thesis in abstract algebra tried a bit to break new ground by axiomatizing exponentials as well as plus and times (I did not anticipate the definition of an exponential as an adjoint functor).

I enjoyed various diversions at Chicago: I learned to play bridge, a standard feature of evening meetings between mathematics faculty and students; I visited the Lyric Opera, but found Wagner's Rhine

Early Years

maidens too buxom; and my friend Manson Benedict and I invented a game of three-dimensional chess. I kept a minor interest in political activities in Chicago. In the spring of 1931, I went along with Manson Benedict and his date Dorothy Jones, a graduate student in economics from Arkansas, to search for a communist meeting in downtown Chicago led by the survivors of the Haymarket Massacre of the late 1880s. We failed to find it, but I later dated Dorothy, as will appear.

At that time, President Hutchins was in deep controversy with many members of the senior Chicago faculty. He had appointed his friend Mortimer Adler as an associate professor of philosophy without consulting the eminent members of the Department of Philosophy. This autocratic action inevitably lead to contention. I took part by sitting in (to the disgust of Bliss) on a philosophy course Adler taught. I did not learn much then, but I did subsequently keep in contact with Professor Adler and read several of his books. Adler was, and continued to be, adept at explaining philosophy to the educated public. He and Hutchins then taught the Great Books and Great Ideas to Chicago students. These topics did not then seem to me to relate to my earlier search for universal knowledge; these "greats" seemed too much constrained by history. However, Great Books and Great Ideas became an important element of the Chicago model of general education.

Overall, I found this year of graduate work at Chicago disappointing, especially because I could not see any possibility for a Ph.D. thesis on logic. I was disappointed with other aspects of mathematics at Chicago as well: there was too much emphasis on pushing students to complete a thesis, which was considered a one-time research experience rather than a preparation for future research. The choice of topics for theses was too restricted to subjects traditionally active at Chicago, such as calculus of variations, matrix algebra, number theory, and projective differential geometry. This tradition had been emphasized far too long and did not fit my own interests. Perhaps I was confirming Professor Ore's judgement on Chicago mathematics. At any rate, I wrote the Institute for International Education, applied for a fellowship to study in

Chapter Five ~ University of Chicago, 1930

Germany and won an award to do so. I notified President Hutchins, who tried to persuade me to stay in Chicago.

My shift from Chicago to Göttingen had a strong, positive motivation: I wanted to write a thesis on logic. At Chicago, there was no evident faculty member available to direct such a thesis, while Göttingen, with Hilbert and others, was an outstanding center for mathematical logic.

Chapter Six
Germany 1931-33

On the way to study mathematics in Göttingen, I spent six weeks in Berlin, in a program at the Humboldt Universität in Berlin designed to acculturate new students from abroad. I attended classes at the university building on *Unter den Linden*, stayed with a German family, had lunch in cafes and tried generally to learn more about the country. I was already aware of pessimistic books such as the *Untergang des Abendlandes* by Oswald Spengler, and I found out about current politics from a long pamphlet presenting the 27 political parties of Germany—the number 27 did shock me. But the most decisive cultural influence was the German poet Bertold Brecht. I saw his play *Die drei Groschen Oper* and was at once deeply impressed by the level of disillusion, as in the verse which described a passing parade of people, "*Denn die einen sind im Dunkeln und die andern sind im Licht und man siehet die im Lichte, die im Dunkeln sieht man nicht.*"[1] This awareness of suffering and total loss of illusions was new to me. I heard tales of the horrors of the German inflation; I contemplated the complex political situation of all those parties with little understanding. It was wholly unlike the two-party system of the United States. In Chicago I had searched in vain for communists but here in Berlin I could find them in plain sight although I did not anticipate in 1931 what would later develop beyond street battles between communists and Nazi thugs. (The *Horst Wessel* song has a verse "Comrades who were shot by the Red Front or the reaction, march now in spirit in our ranks.")

---

[1] "Some are in the dark and others in the light. One sees clearly those in the light, but one does not see the others."

Early Years

Once settled in Göttingen, I did learn more of student life; in particular I learned about the "color fraternities." The color fraternities (*farbentragende*) had a strong nationalistic bent and believed in the importance of personal honor, in a way that seemed almost comical, and certainly was very elitist. They trained their members in dueling with broadswords. Dueling was officially illegal, but the results (in bandaged faces) were evident each Sunday morning. German professors, most prominently in fields of law and medicine (but rarely of mathematics), who sported dueling scars were much admired by many German ladies. I once even came close to fighting in a duel myself. In the winter of 1933, as I was walking to my room, I came across two kids playing with snowballs; I joined in. Presently one of the kids fired a snowball at a passing student (one with fraternity colors). The snowball did not hit, but the student stormed back, enraged, to beat up the kid. Acting American, I intervened. The student then turned on me, as one perhaps worthy of his attention. He said, "Are you a student?" I answered, "Yes." He thereupon asked for my card (it was common to carry cards, similar to business cards). I fumbled in my pockets, only to declare that I did not have a card with me, although I did have such a card in my room. The student then announced emphatically, "We do not concern ourselves with such people," and turned on his heel. In this way I seemed to have lost my honor (in the German sense). But I did avoid a duel, in which my lack of practice with the broadsword would have been very troublesome. The student was true to his pronouncement; whenever he and I passed each other in Lotze Street—and that happened quite regularly—he held his eyes up in the air. Later I discovered that the famous mathematician George Polya was once banished from Göttingen for declining a challenge to a duel.

But I am getting ahead of my story about the university. In Göttingen I found a great deal of mathematical activity. Although the leading mathematician, David Hilbert, had retired and came only once a week to lecture on matters of general intellectual interest—for instance, on the overall importance of the discovery of the Americas—many young stars had appeared. One was Hermann Weyl who had recently come from Zürich as Hilbert's successor. The

Chapter Six — Germany 1931-33

breadth and depth of Weyl's insights were impressive; Riemann surfaces, relativity, logic, and group representations were among the topics he featured. I took part in his seminars on group representations; there I gave a talk about the *Elementarteiler Theorie*, the theory of the elementary divisors of a matrix, which I then barely understood. I even mispronounced the title. I did consult Professor Weyl as to what I should study. He strongly recommended Emmy Noether— saying that she was "the equal of each of us." I did listen to her lectures on linear associative algebras. She made frequent reference to Dickson's influential book (*Algebras and Their Number Theory*), which had just recently been translated into German. I had not heard much about this in Chicago because Dickson was then thinking just about Waring's problem. Noether's lectures were enthusiastic, but often hard to follow because in these lectures she was finalizing the paper she was in the process of writing up. Her power of concentration was impressive—she lectured for three hours, with one break. During the break Paul Bernays and I walked up and down the corridor, discussing logic.

One day when the Mathematical Institute was shut down for an official holiday, Noether announced that we should not give up mathematics for that reason—so we met at the closed doors of the Institute and proceeded by path and fence to climb the nearby hill up to the coffee house at Gerstlingerode Feld, talking mathematics en route, with coffee, and on return. Noether delighted to hear about mathematics, and delighted to so encourage students.

It appears that some years before, Hilbert had also encouraged such hiking to the coffee house. One such hike presented an opportunity to the auditors in his lectures who had noticed that there was a rip in the seat of his trousers. Of course they had not dared mention such imperfection to him. But during the hike they all had to climb over a barbed-wire fence—so one of the assistants rushed up to Hilbert after the fence, to say, "Herr Geheimrat, it must just have happened, but there is a rip in your pants." "No," Hilbert answered. "That rip has been there for a long time." Many such stories about Hilbert were collected and repeated as a tribute to his concentration on mathematics.

Early Years

Edmund Landau, for whom I had wielded a sponge back in Chicago, was very much present in Göttingen. My confreres explained the social life to me. One should have visiting cards printed, then go Sunday morning to present a card at Landau's sumptuous house, which reflected the wealth of Frau Landau's family who had made money from the discovery of a cure for syphilis. A maid received the card, without further immediate response. About a month later, other students and I were invited to an evening party at Landau's. At these parties there were lots of competitive games. A famous story has been told about the mathematician Bessel-Hagen. Bessel-Hagen was a friend of Carl Ludwig Siegel and not very adept at the social life. One day he received an invitation from Landau to one of his parties but he had already bought tickets to the movies. He asked Siegel for advice and received the following suggestion: "Go buy some exquisite stationary and write a note to Landau explaining your predicament that you have already bought tickets to the movies." Bessel-Hagen followed Siegel's advice and was surprised that he was never asked to a party at Landau's house again.

Landau had written many austere and precise books, so his personal style was well known; *Satz ... Beweis* (theorem ... proof), or often just *Beweis klar,* meaning that the proof was obvious. That was the way his lectures were, too—in a big auditorium, with movable (and spongable) blackboards, and supported by a couple of assistants. His current topic was Dirichlet series (not just the zeta function, which is such a series). The properties of these important series were spread out on the board with care and precision, but with no mention of the motivation for the manipulation to which these series were subjected. The Zeta function as an example of a Dirichlet series was hardly noticed. I sat in that big lecture room and took careful notes. Even though there was no motivation, I was pleased to discover that the lecture material was so precise that I was usually able to figure out for myself the probable motivation of his clever maneuvers.

Landau was clearly an expert stylist, as well as an expert on analytic number theory. On that subject, he was often in touch with Professor G. H. Hardy, the leading British expert on analytic number theory. At one point, Hardy was invited to visit; Landau went to the station

Chapter Six ❦ Germany 1931-33

to meet him. As Hardy stepped down from the train, Landau rushed up to him. "Do you have a new method for treating the minor arcs (on the unit circle in the complex plane)?" "No," Hardy said, "I am no longer interested in major and minor arcs." Landau was overwhelmed—until it appeared that his interlocutor was not Hardy but one of the senior Göttingen students masquerading as Hardy. The real Hardy, with the appropriate interests, was soon on hand. Landau enjoyed jokes, even on himself.

Paul Bernays was Hilbert's chief assistant for logic; he was also Jewish, which was not unusual in those days. In the fall of 1931 he was giving the standard Göttingen course, started by Felix Klein and called *Elementare Mathematik vom höheren Standpunkt aus*. It was intended chiefly to train high-school teachers. I took the course because of my interest in logic, but I did not find it exciting. Bernays was also one of Emmy Noether's auditors, so, as mentioned above, in the usual interval between the first and second hour of her lecture, I took the opportunity to walk up and down the corridors with Bernays discussing the latest topics in logic. He had an exceptional knowledge of the developments in the field, as is now apparent from the subsequent book by Hilbert and Bernays on Proof Theory. As was the custom in Germany, the book was based on ideas of Hilbert, expressed in the words of Bernays. Such a cooperation was not considered inappropriate and it was beneficial to the senior professor as well as to his junior counterpart. It represented a different and more authoritarian structure for a university than the American style and has lead to some friction between academics coming from the European tradition and their American students who grew up an in a much less hierarchical tradition.

Note that Hilbert was a universal mathematician, famous for many different accomplishments—in differential equations, algebraic number theory, linear operators, logic, and so on. Bernays was a specialist in the newer field of mathematical logic, and had come to Göttingen as one of Hilbert's assistants. In later years, after leaving Göttingen in 1933, Bernays worked in Zürich. His wide and penetrating understanding of logic became very well-known. On several later occasions I met him and profited from his insights.

Early Years

Gustav Herglotz was another Göttingen faculty member. Herglotz had held the Göttingen chair for applied mathematics for nine years. He was a superb expositor. He held his lectures in the late afternoon so that all the assistants were free to listen. Listen they did, for his lectures were fascinating. All the main ideas came out on the central blackboard, while the necessary calculations were for the sideboards. All was presented in high style. In various semesters, I learned about Lie groups, mathematical optics, and functions with a positive real part from his lectures with these titles. The library of the Institute had copies of his earlier lectures (for example, mechanics), well prepared by admiring students.

Richard Courant was the administrative head of the Mathematical Institute; he managed many things well, continuing the tradition that had been started at Göttingen by that redoubtable organizer, Felix Klein. When Courant was invited to visit the United States, he and his wife invited me to live in his house so that I could help adjust them to speaking English. I am sure I was only a partial success but I did learn quite a lot about speaking German from his children, as well as various pieces of mathematics from Courant. Courant later came to the United States to establish mathematics at New York University.

These were the professors, *ordentlich* (full) and *ausserordentlich* (this rank does not mean extraordinary as the name suggests—it had tenure but did not usually lead to a promotion). As was the practice in German universities, the professors were supplemented by many assistants and *Privatdozenten*. These included many of the outstanding mathematicians of our time, Otto Neugebauer, a historian of mathematics and an editor (then of the *Zentralblatt für Mathematik*), Hans Lewy, Franz Rellich, Richard Cauer, and others. The assistants added a good deal to the depth and richness of mathematics at Göttingen. For example, I learned about partial differential equations from fine lectures by Lewy. Later, after I came to know these assistants better, I often joined them for lunch at the restaurant in the railroad station. Those were opportunities to learn more. To live in Göttengen was to be immersed in mathematical culture.

There were also many active students. Ernst Witt and Oswald Teichmüller were both students of Noether. Witt later developed

Chapter Six ~ Germany 1931-33

many ideas such as the Witt vectors. He was essentially a naive farm boy who took up with the Nazis; it is said that at one point he attended a Noether lecture dressed in his Nazi "brownshirt" uniform. He knew that Noether was Jewish—I'm not sure whether he did this to support the eminent mathematician in front of his peers or to antagonize her. Teichmüller as a student worked with algebras, anticipating the cohomology of groups, but his later work on the Teichmüller spaces made him famous. Fritz John, who later came to the Courant Institute in New York was a student of Courant. In logic I knew Gerhard Gentzen well, and listened to some of his decisive ideas about proof theory. There were many other math students who I did not know well; I also picked up an acquaintance with a non-mathematical student. He took me to a beer-hall party and tried to explain to me some of the aspects of student politics. When I left Göttingen I gave him my bicycle, in exchange for political pamphlets, which he intended to send me — but did not.

I was supported during my first year by a fellowship from the International Institute of Education, an American institute. For the second and third years I got a grant from the Alexander von Humbolt Stiftung, which was a German group that paid for foreign students to come and do graduate work in Germany. It wasn't as big a grant as the previous one had been, but my uncle, who came to visit me briefly in the summer between the two years, gave me a little money to supplement it. I didn't live in such luxury as before; actually, I never did. But between the Humboldt Foundation and my uncle, I managed to scrape up enough money to get by.

The *Mathematische Gesellschaft* met every week to hear a visiting lecturer. Before each lecture, tea was served—students were invited—in a special tea room where the current issues of mathematical journals were on display. The lectures were generally interesting, and on occasion produced stories. Once John von Neumann came to lecture on quantum mechanics, making use of Hilbert spaces. In Hilbert's original work on integral equations, a point in Hilbert space was an infinite sequence of complex numbers y_n with the sum of squares convergent. But von Neumann's lecture began thus: "Take a Hilbert space—an infinite-dimensional vector space over the complex numbers complete in a positive definite norm." At the end of the lecture Hilbert asked Professor von Neumann, "I would like to know, just what a Hilbert Space is." Hilbert thought of his spaces concretely, not axiomatically.

At one point Ludwig von Mises, then a professor in Berlin, came to give a lecture on his new views on probability. For him, a probability space consisted of an infinite sequence of zeros and ones such that the frequency of ones is not altered by any uniformly lawful selection of a subsequence. The entire Göttingen establishment vigorously contested this bold statement. "Dr. von Mises, just what is meant by a lawful selection of a subsequence? Perhaps the subsequence of all Ones?" Dr. von Mises had a hard time conducting his defense. There had been a long-standing competition between the mathematics faculties at Göttingen and at Berlin that came out in this confrontation.

From the contrast between Göttingen and Chicago I learned a great deal. In Chicago (and at Yale) a vector in an n-dimensional vector space (over the real numbers) was an n-tuple $(y_1,..., y_n)$ of such numbers. In Göttingen a vector was an arrow, and a vector space consisted of objects (vectors) which could be suitably added and multiplied by scalars, as in $a(y_1..., y_n) = (ay_1,..., ay_n)$. The corresponding axioms for a vector space were well known; for example, Hermann Weyl's presented them in his famous book on relativity theory, *Space, Time and Matter*. The use of axioms to describe algebraic objects was a basic element in Emmy Noether's view of algebra. In her view, algebra should deal with concepts and not just manipulation. It was only in Göttingen that I came to understand these things well—an understanding that was important to my later exposition of modern algebra in my joint book with Garrett Birkhoff, *A Survey of Modern Algebra*.

There were other new mathematical ideas that I did not fully grasp at the time. Weyl's lectures on group representations pointed out firmly that there should be a study of Lie algebras (such as was later undertaken by Nathan Jacobson). Emil Artin came to visit and to give lectures in class field theory, which, to my regret, were well over my head. I was at one point assigned to write up the notes for Weyl's lectures on the foundation of mathematics; I did not complete this task. Mathematics was blossoming; I did not learn it all.

Chapter Six — Germany 1931-33

I had come to Göttingen intending to work in mathematical logic—and I did. However, I did not then catch up to its newest development: Gödel's famous incompleteness theorem appeared in 1931, but at that time I did not study it carefully—an amazing omission, since Gödel's work was among the most important in 20th century mathematics and a major turning point in logic. I no longer understand why I missed this event. At the time, Gödel was on the faculty of the University of Vienna; I wonder whether his work was suppressed in Göttingen because it was a blow to the work of Hilbert and others, or whether I was just so busy that I ignored the excitement. It was only later, back in the United States, that I studied Gödel's work.

When I told Weyl I wanted to write a thesis in logic, he said, "Well you should go talk to Bernays." So I did talk to Bernays, who was quite accessible and he became my advisor. When I first began to write my dissertation, Bernays strongly disapproved of it. So I threw it away and wrote another.

In 1932, I apparently thought that a Ph.D. thesis should be something brand-new and different. I did not grasp the idea that mathematics is like a building whose elements support one another. I was still much impressed with the important contribution of Whitehead and Russell in formalizing mathematical proof. This became the central point of my actual Ph.D. thesis, "Abbreviated Proofs in Logic Calculus." It was printed in Göttingen in 1934 by Hubert & Co., and later reprinted in a volume of my "Selected Papers."

My thesis plan was approximately as follows: imagine all mathematical proofs written in the formal style of Whitehead and Russell. It will then be apparent that there are several common shapes or plans for a proof, for example, the scheme such as using Modus Ponens to combine Theorems 2.8 and 4.7. Make tables and lists of all of these common schemes; this will provide a way of abbreviating the proofs, which are thereby mechanized as combinations of specified common patterns. Once all proofs are done so, one has at hand a classification of types of proof and methods to abbreviate each proof. This is what my thesis tried to organize.

My thesis did not attract any following, nor did it have any influence on subsequent studies of mechanized proof. It was naive in supposing that all mathematics would be *actually* written in Whitehead and Russell style. Nothing of the sort happened. My thesis was accepted, with the grade Genügend—the lowest grade. Neither Paul Bernays nor Hermann Weyl were impressed by it.

Dorothy Jones, who I had met in Chicago, had come to visit me in Göttingen—and Europe generally—in the summer of 1932. We visited Vienna, Budapest, and Venice. In 1933 she came over again to help me finish up (and type) my thesis. We decided to be married—after I completed my doctoral exams. It required some bureaucratic preparations; for example, Dorothy had to make her intentions public in a German newspaper in New York City, where she had been living.

In the meantime, politics had become dominant. In February 1933 there was a decisive election in which the Nazi party, combined with the conservative party (the Deutsch Nationale Partei), won a majority of the seats in the Parliament. Hitler then became chancellor. There began a drastic change that soon deeply affected universities, and mathematics in particular. At the end of February, I had gone off on a student ski trip to the Tyrol. As our party came back on the train, we read the papers to learn that there was to be that day a boycott of all Jewish-owned stores. Our train stopped for a couple of hours in Nürnberg, so we all left our skiing gear in the train and went off to see the town. There I saw a prosperous shoe store, closed, with a seedy-looking man inspecting the show window. Presently, the police came by and hustled the man off. Until then, I believed what I had read in the paper—that the boycott was to be peaceful. This made me curious, so I followed after the seedy man and his police escort. When we got to the police station, I, too, was taken into custody. I was clearly not a German, so the police officers thought I must be one of those foreign reporters purported to be writing false accusations about Germany in the British paper. I tried my best to convince them that I was only a student and not a reporter—and that all my belongings were on a train that would soon leave for Göttingen! The police did not seem to believe me, but they did let me go, just in time to catch my train.

When I got back, I went to my usual room—rented in the apartment of a German widow. She regularly prepared an evening

Chapter Six ~ Germany 1931-33

meal for me, and she enjoyed talking. But this time, to my surprise, her talk was full of praise for Hitler and his plans. In my two-week absence she had regularly read the papers and so had become converted to the Nazi ideas. It was a striking example of the power of unrestricted propaganda. I did not really try to change her opinion—my individual views would not have mattered in the flood of Nazi publicity. But I did not tell her that I had a copy of Marx's *Das Kapital*; instead I hid it under my shirts in the drawer. It was then saved from burning, the fate of many such books in Göttingen.

Soon after coming into office, Hitler issued regulations about the faculties of universities. All the universities were funded by the state, so all the faculty and staff were government employees. The new rule stated that most Jewish faculty members must be immediately dismissed. There were some exceptions, for those faculty members who had been in office for 25 years and for those who had fought in the German army in World War I. This had the following effect on the professors of mathematics in Göttingen:

Professor Richard Courant was immediately dismissed, despite the fact that he had served in the German army and had been wounded in World War I. Courant, as administrator of the Mathematical Institute, had been very influential in the German mathematical community at large and was suspected of being a socialist.

Professor Edmund Landau, although Jewish, was left in office; he had been a full professor for over 20 years and had served in World War I. However, in the following winter semester, his lecture on calculus was boycotted by the Nazi students. He thereupon resigned.

Emmy Noether, though probably the most important woman mathematician of all time, was Jewish and was suspected of left-wing sympathies, so she was dismissed. She found refuge (for a brief two-year period before her death) at Bryn Mawr College in the United States.

Paul Bernays, a Jew, was immediately put on leave of absence and subsequently dismissed. For me, this meant that I had to find a different thesis advisor. Bernays later took refuge at the Eidgenössische Technische Hochschule in Zürich.

Early Years

After Courant's dismisal, Otto Neugebauer was appointed Director of the Mathematical Institute. However, although he was not Jewish, he had liberal political views and was suspected of left-wing sympathies. He lasted only one day as director.

Hans Lewy and other *Privatdozenten*, many of them Jewish, were dismissed.

Professors Herglotz and Weyl (neither Jewish) were not dismissed. However Weyl's wife was Jewish, so his two sons would count as Jewish. He had recently turned down an offer of a professorship at the recently established Institute for Advanced Study in Princeton. Weyl apparently requested that the offer be renewed; it was, so he left Göttingen at the end of the spring semester.

The net result of all this was that only one (Herglotz) of the four full professors was left in office, and many of the junior faculty were dismissed. In effect, the Nazi anti-Semitic rules had been used to ruin the Mathematical Institute at Göttingen. It was a very sad denouement. Prior to 1933, the atmosphere at Göttingen crackled with enthusiasm for mathematical ideas and discoveries. It was a remarkable exhibit of the university as an amalgam of research, teaching, and inspiration. But the glory of mathematics at Göttingen came to an abrupt end when Hitler came to power. The Nazi government wrecked what had been the most prominent and influential center of mathematics in the world. The total effect has been described with great care in a recent volume, *On the University of Göttingen under National Socialism*, by Heinrich Becker, Hans-Jopachim Cahms, and Cornelia Wegeler (1987). But no one volume can convey the force of the resulting devastation of scholarship. Many of the individuals concerned did find positions in other countries (Weyl, Noether, Courant, Lewy and others in the United States; Bernays in Zürich). But the magical place, Göttingen, was destroyed.

There was a great deal of confusion. We students hastened to finish up and get out—as for example, in the case of my thesis. When Bernays disapproved of my first thesis, I pressed very hard to get the second one done. Since Bernays was not officially in office anymore, the thesis was submitted to Hermann Weyl. Weyl evidently didn't think it was very good, so it got the lowest passing grade. But to earn my degree, there was not only the rush to finish the thesis. There were also final oral exams to be taken on the major and two minors. Weyl conducted the oral exam on the major. This must have been for Germans a very severe process, because they had no examinations for the six or seven years they studied. But with an American background, I had taken all sorts of examinations and I crammed for this examination with Weyl, studying just what he told me to study. When Weyl came to the exam, I noticed that he had prepared a long list of questions. I decided right there that I would do better if I could get him to stick to the questions he had prepared in advance than to let him ask me something spontaneously, so I made sure he did not get to the end of the list—I explained every answer in great detail so he didn't have time to ask anything more. This way, he was impressed by my knowledge and he gave me an *ausgezeichnet*—the highest grade—on the examination. To some extent, that balanced out the *genügend* on the thesis.

My minor exams were with Professors Herglotz and Geiger, the latter in the philosophy of mathematics. I prepared myself for Geiger's test by listening to his lectures on the subject. He was evidently Jewish, but had served in WWI, so was not immediately dismissed. I can distinctly remember going to his lectures and being amazed at how nervous he was. It was obvious that he was under a great deal of pressure from the threat of the Nazis and that he was badly worried by his prospects.

The coming oral exam with Professor Herglotz worried me — he was such an austere and knowledgeable professor. So I consulted some of my friends. They told me to remember, he likes to lecture. When the exam came, his first question was: "Well, Mac Lane, do you know Felix Klein's Erlanger Programm?" I did; it asserted that each geometry is determined and characterized by its group. Then he asked, "In the case of complex analysis, which is the group?" I managed to say that it was the conformal group. With these answers, Herglotz proceeded to give an elegant 30-minute lecture describing complex analysis in terms of this group. I was given no additional questions to answer.

Early Years

Saunders Mac Lane and Dorothy Jones Mac Lane on their wedding day, July 21, 1933

On July 21, two days after I took my doctoral exams, Dorothy and I were married in the City Hall at Göttingen. It seems that afterward I went to hear some mathematical lecture, but later we did have a small banquet (with two guests) in the Ratskeller.

When Dorothy and I went to the Standesamt to get a marriage license, we were surprised to find my mathematical friend Fritz John there with his girlfriend, Charlotte, who was also a student of mathematics. He was Jewish, she was not, so they were hastening to get married before such marriages were ruled out. They wanted to keep their marriage as quiet as they could, so they swore Dorothy and me to secrecy and asked us to be witnesses at their marriage. After the wedding, they left Germany on separate paths and were reunited in New York.

With exam, marriage certificate, and thesis all in hand, Dorothy and I then left Göttingen, and Germany, for Paris, traveling by train and stopping at Heidelberg en route. I was convinced that the Nazi rule in Germany would lead to no good end and indeed would probably precipitate another World War. It does not help that I was right.

Once I actually saw Hitler. While in Göttingen it suddenly occurred to me that I had never visited Weimar, the famous court where

Chapter Six ⁓ Germany 1931-33

Goethe lived and worked. So I took the train and found a hotel, then went to the opera house. The tickets for the Wagnerian concert that evening were all sold out—it was Wagner's birthday, May 22— but someone with an extra ticket came along and sold it to me. I went and enjoyed the opera, and went out into the lobby for the intermission. There, across the lobby, stood Hitler and Göring—they were easy to recognize from the many pictures I had seen in the papers, and it was well known that Hitler liked Wagnerian opera. I had no gun with me.

Göttingen mathematics, before Hitler, was clearly a rousing success. The Mathematics Institute there had long aspired to represent all the current active developments in mathematics—and so it was in my time: Lie groups, quantum mechanics, logic, analysis, number theory, abstract algebra, partial differential equations. A student in Göttingen could be aware of all this, and could specialize as he or she wished. To be sure, Göttingen had acted to take over fields (for example, integral equations) that had started elsewhere. This policy even had a name: "nostrification"—making it ours. For those studying in mathematics at Göttingen, this did provide it all. What an experience, and what a loss at the end.

As a supplement to this account of my mathematical studies in Göttingen, I append excerpts from letters that I wrote at the time. The first describes my estimate of the political situation as I saw it then:

> There are times when even the most serious and important of events can be laughable. One can't deny that important events are important, and here in Germany the revolutionary ideology and the amazing dynamics of the National Socialist Movement are certainly highly important. But at the same time the self-dignity of important events does tend to the ridiculous—and it is laughable to hear all the Nazi papers howl because they're getting a dose of their own medicine in Austria. All the well-tried Hitler methods of eradicating Marxism— imprisonments, press censorship, newspaper Verbote, Verbote of opposing parties, and the like—are being used against Hitler's cohorts in Vienna.
>
> The Nazis naturally emphatically proclaim the great differences between the two cases of oppression. In Austria, you see, the Hitler

movement is anchored firmly in the Volk, hence any autocratic attack on the movement is a defamation of the sacred rights of the Volk. In Germany, on the other hand, Marxism and Communism, and liberalism and Judaism are pests and blights upon the Volk, are antiquated ideas that must be crushed if the Volk is to be able to live on, and are dangers that must be warded off if civilization is to be saved from destruction (*Untergang des Abendlandes*). The Volk, the salvation of the Volk, and the progress of the Volk are everything.

This is only one example of the ways in which the Volk is made the basis of everything in present-day Germany. The culture is determined by the Volk and so must be rooted in the Volk. Science, industry, and religion are servants of the Volk. And so on, always in the same pattern. The statements one hears so frequently in Germany that they become nauseating—especially since they are never subjected to any criticism whatever.

But underneath the monotonous repetitions, phrases, and these bombastic-mystic proclamations lies a fundamental conflict; a conflict between the individualistic-rationalistic liberalism of the Anglo-Saxon and the racial and collectivistic authoritarianism of the Prussian. It is a basic conflict, for it expresses itself in many minor conflicts; in pacifism versus militarism; in absolute freedom versus freedom to serve the Volk; in national culture versus world civilization; in parliamentarism versus dictatorship. IT MAY WELL BE THAT THIS CONFLICT WILL IN ITS MANY PHASES BE ONE OF THE DOMINATING FEATURES OF THE 20th CENTURY.

The fundamental character of the conflict is well-illustrated by the fury with which it upsets everything, even the most stable and aloof institutions. Thus it is that the German universities are undergoing a tremendous reorganization. The Göttingen mathematics department, once one of the best in the world, is now but a skeleton. Professors have been fired, professors have been given leaves of absence, and the few professors who remain are desolated and depressed by the attacks. And this is not chance, for the dynamic basis of a people's revolution is opposed to such an abstract, intellectual, specialized, and international science as mathematics. Such a science is characteristic of the "liberal" civilization that the Nazis wish to destroy.

Chapter Six ～ Germany 1931-33

But there is no lack of proclamations as to how the University is to be reconstituted in the Nazi Reich. Everything must serve the Volk, and so the University, too, must be rooted in the Volk. The students must not be simply Academiker, they must keep contact with the Volk and with the working class. They must be schooled in a spirit of soldierly discipline and responsibility. Thus it is that the Nazis hope to restore a coordinated significance to the specialized abstruseness of the modern universities.

A second item concerns the writing of my thesis. As already noted, I chose my thesis topic on my own, although it was then and now customary to have a thesis topic assigned or suggested by the professor. Perhaps for this reason I had worried from time to time, but mostly I felt excitement, as indicated in the following excerpts from my letters to my mother:

January 23, 1932

And in the midst of all this whirl of social activity (a dancing party with Professor Weyl and a dinner party with Professor Landau), I am finding great ideas and magnificent generalizations of my theory of logical and algebraic systems in general. I have really enjoyed these ideas tremendously—they are so gloriously general, they suggest all sorts of vague possibilities of application to other fields of knowledge. I don't believe I have ever enjoyed myself so thoroughly as in following them up and thinking them out. To work them out will take a long time, but that does not matter.

December 20, 1932

At present, most of my time is still taken up with finishing the first two parts of my thesis. The first part, consisting of some 80 typewritten pages (not as much as it sounds, for it is triple-spaced and has one third of the page as margin) is now done and has gone to German students to be grammatically investigated. I hope to get this paper to Professor Weyl by the end of the week. The rest may be finished up in vacation.

Early Years

February 8, 1933

For a time I had gotten thoroughly disgusted with the intolerance, the narrow-mindedness, and the utter lack of philosophic grasp displayed in the attitude of the hiesige (local) professors toward my thesis. On Monday I was prepared to transfer to the University of Vienna, where there are many who think as I do. But just then I had a conference with Professor Bernays, told him of the philosophical aim of my thesis and succeeded in persuading him that it was possible as a doctor thesis. Immediately thereafter I got a new idea (goodness knew I already had ideas enough to fill up any three theses as soon as I got a chance to write them out). But this particular new idea has certain practical advantages; it happens to be the kind of idea that the professors here think is the sine qua non of a thesis. I don't think it is much more important than other ideas, but they do. Hence I should be able to further pacify them by writing up this particular idea and putting it in this thesis.

April 17, 1933

I have had a tremendous brainstorm this past week—I have been almost oblivious of time and space. I have come out with an idea that, for all that I can see, is absolutely a worldbeater. It may have tremendous general significance but at any rate it is absolutely certain to settle the degree question—and that right soon.

April 18-22, 1933

The title page of my six-volume "Collected Works of Goethe," purchased at that time, carries the inscription, "In celebration of the discovery of the logic calculus 18IV33 to 22IV33, meaning that I "discovered" the concept of a logic calculus between April 18 and April 22 of 1933.

April 20, 1933

Perhaps I have time to tell you about my new discovery. It is a new symbolic logic for mathematical proof. It applies, as far as I can see, to all proofs in all branches of mathematics (a rather large order). It makes it possible to write down the proof of a theorem in a very much shorter space than by the usual method and at the same time it makes the proof of the theorem very much clearer. In essence, it eliminates practically all the long mechanical manipulations necessary to prove a theorem. It is only necessary to give in sequence the leading ideas of the proof. In fact, once these leading ideas are given—together with a few directions—then it becomes possible to compute from the leading idea just what the proof of the theorem will be. In other words, once the leading ideas are given, all the rest is a purely mechanical sort of job. It is possible to define once and for all how the job is to be carried out. (This generality definitely depends upon the abstract method I have been developing for the past year.) Since the execution is purely mechanical, there is no use in carrying it out in detail in every proof. In essence, this detail is no more important than the detail of a multiplication such as $25 \times 25 = 625$.

May 28, 1933

As for the thesis, the progress is OK.

June 8, 1933

The thesis is now in its very last stage and so requires extra effort. Weyl has already seen half of it. This time he seems inclined to accept it.

June 15, 1933

As to the thesis, it is now done, all 125 pages of it. Weyl is now engaged in reading it. In a few days I will be able to tell you of his reaction to it. But whatever his reaction may be, I shall have the definite feeling that I am on the track of something very important and that this track can give me a lot of fun in the next few years.

In retrospect, my thesis had no effect, although it was a very early essay in the subject now known as automatic theorem proving. It was published, as all German theses were, in 50 copies that I distributed to various of my friends. It was in German, however, so none of them paid much attention to it. It was republished years later in the selected papers of mine, but it still has had little effect on further work.

Early Years

At that time, mathematical logic was only just beginning to be a respected field in mathematics. Most mathematicians thought it belonged in philosophy, so it was hard to get ahead or get jobs in it. For that reason, shortly after the thesis, I shifted to work in algebra.

Part Two

First Teaching

Chapter Seven — Yale and Harvard

By September 1933, Dorothy and I were back in the United States. My friends at Yale (probably E. J. Miles) providentially arranged for me to do postdoctoral work as a Sterling Fellow for the year 1933-34. My stipend was $1,000 a year, and Dorothy soon found a job researching automobile traffic that paid $300; we lived on these combined earnings. We rented a one-room apartment with a rent of $35 a month, bought some necessary furniture, and settled down. We were greatly helped by the fact that I received free meals as a faculty fellow.

I had an additional appointment as a Faculty Fellow of Jonathan Edward College, which was part of Yale's new plan for all undergraduates to live in residential colleges. I enjoyed pleasant times with the other faculty fellows. We often shared dinners; at one point, I was instructed to serve the sherry after dinner, but I served the port instead—I didn't know the difference because my father was a minister. But I was not much harmed by my mistake: The practice of serving sherry was just a small effort to align Yale habits with the traditions of the senior common rooms of the colleges at Oxford and Cambridge.

Professor Ore supervised postdoctoral fellows in mathematics; hence, as the only fellow that year, I was expected to do research in algebra, which I did with considerable enthusiasm. I recalled easily the excitement of the courses I had taken with him as an undergraduate. Ore's research in algebraic number theory included a constructive study of the factorization of a rational prime number p in an algebraic number field (that is, a finite extension K of the field Q of rational numbers). Every ordinary integer m has a unique representation as a product $m = p_1^{e_1}...p_n^{e_n}$ of prime integers, but unique factorization fails for other domains of algebraic integers such as $a + b\sqrt{5}$, where there is no such decomposition into primes. However, Richard Dedekind, a 19th century algebraist, formulated a decisive first example of "abstract" algebra: In such cases, there could be a unique decomposition into new objects—the prime ideals. Thus, in each such field K, any rational prime p has a unique presentation $p = p_1^{e_1}...p_n^{e_n}$, with integral exponents e_i applied to prime ideals p_i of the ring of algebraic integers of K. Emmy Noether used the concept of an ideal for more general rings.

Subsequently, Noether's students, and Ore, were able to construct prime ideals in a different way: as functions called valuations. For example, if p is a prime number in the ring **Z** of ordinary integers, then each integer m has a unique factorization as $m = p^e m'$, with e an integer and m' relatively prime to p. The resulting function V_p defined by $V_p(m) = e$ has the following readily established formal properties: $V(mn) = V(m) + V(n)$, and $V(m + n)$ is greater than or equal to $Min(V(m), V(n))$, with $V(0)$ undefined. Such a function V is called a valuation of the ring **Z**, while two valuations V and V' of **Z** are said to be equivalent when $V' = kV$ for a constant $k \neq 0$. It is easy to verify that the equivalence classes of valuations of **Z** correspond exactly to the prime numbers p of **Z**. Similarly, Dedekind's prime ideals p_i correspond exactly to the valuations of the associated ring of integers. In other words, Ore's constructive approach to algebraic number theory could be described in terms of valuations. I carried this valuation theory further in a series of papers through explicit construction of all the valuations of certain fields and their application to classical algebraic number theory, and to certain irreducibility criteria.

In Germany, Wolfgang Krull (a student of Noether) introduced valuations V of high rank when V takes values not in the integers, but in a general ordered abelian group. I studied these generalized valuations upon Ore's suggestion; they had (as Krull knew) possible applications to algebraic geometry, where a valuation could represent the multiplicity of a function at a singular point of an algebraic variety. At the time, I was only vaguely aware of this use of higher valuations (I did not know enough algebraic geometry). Oscar Zariski, who was familiar with Italian algebraic geometry, later developed the idea effectively.

There is a close relation between algebraic numbers and algebraic surfaces: an equation

$$a_n x^n + a_{n-1} x^{n-1} + \ldots + a_1 x + a_0 = 0$$

has the algebraic numbers as roots when the a_i are rational numbers. If the a_i are polynomial functions of y, the equation describes the points (x, y) on the algebraic curve. At that time, algebraic number theory had been systematically developed in Germany. Algebraic curves and surfaces had been extensively studied in Italy, with more emphasis on intuition than rigor. With my then firm belief in the importance of explicit rigor, it was difficult for me to understand the results of the Italian algebraic geometers.

During my research on valuations, I began to inquire about possible teaching positions for the upcoming year. I found nothing available at universities, but I did locate a prospective position as Master at Exeter, one of the prep schools that had seemed so intimidating to me as a freshman at Yale. I joined the American Mathematical Society in hopes that it would aid in my job search—I attended my first meeting in Cambridge, Massachusetts, in December 1933. As was the custom, I gave a 10-minute research report; I chose to speak about an aspect of my thesis on logic. Professor Ore probably shared the then common opinion that logic was not a fully respectable part of mathematics, or perhaps he was disappointed that I had not chosen algebra as my topic; whatever the reason, following the usual call for discussion after a talk, Ore rose to denounce my work. He did so in a large general session, and I was aware that most of the Harvard mathematics department were present to hear both my paper and Ore's comment on it.

However, early in February 1934, Professor William Graustein, Chairman of Mathematics at Harvard, wrote to offer me a one-year appointment as a Benjamin Peirce Instructor, with probable

First Teaching

The apartment on Harvard Street, 1934-1936

reappointment for a second year. Apparently, my experience with Ore did not prevent me from getting a job. The Benjamin Peirce Instructorships, named for an eminent 19th century Harvard professor, provided temporary teaching positions with low teaching loads to help young research mathematicians get started (the idea was, and is, widely imitated by other math departments). I accepted the offer at once, and withdrew my application at Exeter. The salary at Harvard was $2,250—I felt like a rich man. Dorothy and I upgraded to a comfortable three-bedroom apartment, and we boldly invited Professor and Mrs. Graustein to dinner.

Professor Graustein told me that, in addition to teaching calculus, I would give a half course on a topic of my choosing. I responded that I would like to teach a course in mathematical logic, but that I could alternatively give a course in algebra. Graustein indicated that the department would particularly value a beginning graduate course on algebra, as the only faculty members working in algebra had recently left. So I taught a half course using van der Waerden's recent two-volume *Modern Algebra* as the text: it was a fine book, presenting Noether's modern conceptual (or abstract) view of algebra. This course was an excellent way to start teaching algebra; my students were very able. I also enjoyed teaching calculus to freshmen; of course, I used the standard

Chapter Seven — Yale and Harvard

Harvard textbook by William Fogg Osgood (inevitably deemed "Foggy" Osgood), who had recently retired from his professorship.

In addition to teaching, I enjoyed working as a tutor. Harvard's new tutorial plan for undergraduates assigned all students above freshman level to a faculty tutor in his major: I advised ten undergraduate mathematics majors, meeting with each of them once a week. Students came to my office in Eliot House for advising—Harvard was also developing a residential house system for students that seemed to be functioning quite well; perhaps better than the corresponding system at Yale.

The mathematics department was lively, even though there was not yet any central administration—there was not even a department office. A few favored faculty did have offices in the stacks of Widener Library. George David Birkhoff was a dynamic but unofficial leader; he actively attended to visiting fellows such as Magnus Hestenes (from Chicago). Other faculty included Lars Ahlfors, Julian Lowell Coolidge, Graustein, E. V. Huntington, Marston Morse, Joseph Walsh, Marshall Stone, and David Widder. Ahlfors, from Finland, was a profound student of analytic functions; Coolidge, a geometer, was a member of the New England aristocracy; and Huntington was a venerable student of classical axiomatics, especially Boolean algebras, and he also taught statistics (for example, how to apportion congressional districts). Morse had a particular enthusiasm for his own ideas: I remember standing about street corners listening to him speak about his newest interests, such as the calculus of variations in the large. Osgood married Morse's ex-wife and left town with her; Morse suffered for this, which was publicly recorded in the ditty:

Here's to Marston, Mickey Morse
A man experienced in divorce.
His opinion of himself, we charge
Like nose and book is in the large.

There is no doubt that he liked to hear himself talk, but in my experience, Morse was a real stimulus to all who listened.

First Teaching

Marshall Stone was a student of George Birkhoff, with a thesis on differential equations; for example, $ay'' - by' - cy - f(x) = 0$. In such equations, the functions y and f of x can be regarded as points in an infinite-dimensional space, known as a Hilbert space. Simultaneous to Stone's studies in Cambridge, von Neumann was exploring these spaces in Berlin. When von Neumann sent one of his manuscripts to Stone, Stone noticed that the classical notion of an adjoint differential equation applied, yielding adjoint operators in the space. He wrote to von Neumann, who saw this point and wished to modify his papers, which were already set in type, ready to print. His publisher, Springer, agreed to make the changes only if he agreed to write a book on the application of Hilbert spaces to quantum mechanics, a book von Neumann prepared in 1932. The same year, Stone published a big book on linear transformation in Hilbert space.

Stone later studied Boolean algebras, which axiomatized the logical operations "and," "or," and "not." He showed that "and" and "or" could serve as multiplication and addition for a ring, thus showing that the Boolean algebras (as axiomatized by Huntington and others) fitted well with abstract algebra (rings) in the sense of Noether's abstract algebra. His idea was that mathematics needed thought rather than calculations. At one point, Stone and his family lived in one branch of a U-shaped apartment building. A university administrator living in the opposite branch watched Stone at his desk one evening, and was amazed to see that Stone wrote nothing—he just sat there thinking. Stone represented the mathematics of ideas, not of calculations.

Senior faculty members visited sections of undergraduate courses taught by young instructors—Coolidge visited my course. His only comment on my teaching was that for 40 years he had been talking too fast in his classes and he could see that I was continuing this practice. I did not remind him of the popular story that during class one day, while twirling his watch on its chain, the chain broke sending the watch out the window. His response: "Ah gentlemen, a perfect parabola." As I mentioned, he taught geometry.

Hassler Whitney was an assistant professor. As a former student of George Birkhoff, much of his research furthered Birkhoff's insights

Chapter Seven ❦ Yale and Harvard

in algebraic topology, measuring the connectivity of spaces by Betti numbers (homology groups). Whitney's later research in topology amounted to decisive contributions such as the fiber bundle, in which the bundle consists of all the tangent lines at all the points on a given surface. During the time I was at Harvard, Whitney was thinking about the four-color problem: Can any map of countries on a sphere be colored with at most four colors such that no two adjacent countries have the same color? Examples suggested this was possible, but at that time no one could find the proof.

A component of this problem was to determine if a proposed map can actually be drawn on a sphere. There is a famous example where the map cannot be drawn: consider five embassies in Bern, Switzerland during the war. Each embassy wishes to communicate with every other embassy by a tunnel. Because of a rocky foundation, each tunnel must be the same depth underground (say exactly 12 feet), and no two tunnels may meet. Before digging, consider the situation—it can't be done. In the attempt below, there can be no tunnel between A and D.

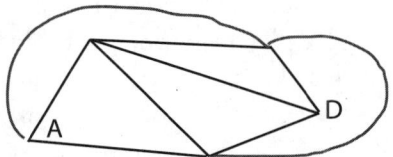

There was another troublesome example: suppose that in some city, three utilities (gas, oil, and telephone) represented by points A, B, and C below wish to communicate with three customers represented by points D, E, and F with nonintersecting pipes, each a fixed depth underground (say 8 feet). Try it—there is no way to get from B to E.

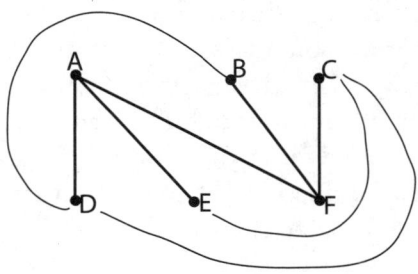

First Teaching

In these two examples, the graphs could not be drawn in the plane without crossings. The mathematical problem was to characterize those graphs that can be drawn in the plane (equivalently, on a sphere) without any extra crossings. This problem led to a minor competition. At the University of Michigan, the topologist W. L. Ayres tried to prove that every non-planar graph must contain a copy of the 3 × 3 graph in the second example. At the same time, the famous Polish topologist C. Kuratowski tried to prove that every non-planar graph contained a copy of the (diplomatic) complete 5-point graph in the first example. At an international conference, Ayres and Kuratowski met and compared examples. With two examples already, Ayres decided that there must be many more examples of non-planar graphs, and so gave up. Kuratowski, however, went home and proved that every non-planar graph must include *either* the 3 × 3 or the complete 5-point graph.

Whitney's interest in these matters led him to a different theorem: a graph is planar if and only if it has a dual. To construct a dual to a planar graph, put a new point in each region and join any two new points in adjacent regions. I found this result interesting, and discovered a variant: in a planar graph, the boundary of each region is a circuit (a simple closed curve). The boundaries of all the regions form a collection of circuits that contain every edge of the graph exactly twice, and these bounding circuits generate all possible circuits. I published this theorem in the Polish journal *Fundamenta Mathematica* in 1936. It was fun to do research outside of algebra and logic, and I learned a great deal of topology from Whitney.

I learned about universal algebra from Garrett Birkhoff (George Birkhoff's son), who was one of Harvard's new "junior fellows," a group established by President Lowell so that there was a place for advanced students who couldn't be bothered to get a Ph.D. This degree was controversial; some saw it as a sort of octopus, strangling serious graduate study. Willard Van Orman Quine was also a junior fellow. As an undergraduate at Oberlin he was much impressed with *Principia Mathematica* and came to Harvard to work with Whitehead, who moved from England to join the philosophy department. Quine rewrote *Principia* and became a leading logician in the department.

Chapter Seven ~ Yale and Harvard

Overall, I did not lack for people with common interests in logic. For example, I met Barkeley Rosser, who studied at Princeton and came to Harvard as a postdoc. I also knew and liked his good friend Steve Kleene, who also came from Princeton. But I made no real contribution to mathematical logic then; the ideas from my thesis were left dormant. However, I did join the newly established Association for Symbolic Logic.

I had other research activities in topology. Dorothy grew up in Fayetteville, Arkansas, where her father, Virgil L. Jones, was a professor of English and dean at the University of Arkansas. I met Mr. and Mrs. Jones in 1933, just after our return from Germany. We later visited in Arkansas and met Dorothy's two younger sisters, Isabel and Alice. While there, I looked to find mathematics, and so met V. W. Adkisson at the university. He studied an "extendability" problem in his thesis, directed by J. R. Kline at the University of Pennsylvania. Given a graph G on the sphere, when can a homeomorphism mapping G to itself be extended to the whole sphere? Working with earlier results, we found a new condition that every homeomorphism $G \to G$ can be extended in every map of G on a sphere, and we found useful variants of this new condition. In the course of this work, we had occasion to use a 1934 result of Samuel Eilenberg on periodic transformations of the sphere. I did not know Eilenberg at the time, but in 1940 I met him when I lectured on our result at a topology conference at Michigan. My interest in topology, later of considerable scope, developed from several accidents; studying Hausdorff's book at Yale under W. A. Wilson, listening to Whitney at Harvard, and then working on graphs with Adkisson at Arkansas. The decisive joint research in topology with Eilenberg was still to come.

In my first year at Harvard, I received a letter from General Electric's research laboratory (where I had previously worked with Irving Langmuir in college) proposing a summer appointment for me in 1935. I realized that this might well open up a career for me as an industrial mathematician, but after some thought, I declined the offer choosing instead to spend the summer on my own research. My decision amounted to a conscious choice in favor of a university

First Teaching

career, which was a gamble—university jobs were scarce during the Depression.

After Christmas in 1935, I went to the American Mathematical Society meeting, again looking for a teaching position. There were several job openings at Cornell that year; so many of us applied that Wally Hurwitz, the Cornell Chairman, conducted a group interview with all of us in a beer parlor. Several of us won jobs—Barkeley Rosser, Robert Walker, John Curtis, Dan Lewis, and John Randolph are some. There were only a few other jobs that year; Magnus Hestenes, who had just spent the year on fellowship at Harvard, took a position at UCLA.

Dorothy and I left Cambridge grateful for the two stimulating years there, and pleased that I had a job in hand. The time at Harvard provided a good start for both teaching and research; the presence of an eminent faculty mattered greatly.

Chapter Eight
Cornell and Chicago

In 1936, when I arrived at Cornell, all of the new instructors (and some more-experienced faculty members) were assigned to teach freshman calculus. Since we were all in the midst of a competition for permanent positions, we decided to give a common final exam so that we could compare our teaching successes. I recall that my section scored the highest on that exam; for a moment, I felt that the scores were the effect of my splendid teaching—how fine—but then I discovered that my section was scheduled at a time convenient for those engineering students who would go on to study chemical engineering. It was well-known at Cornell that the chemical engineers were the best of the engineering students, so my section's success on that exam was probably independent of my teaching—so much for our competition.

I also taught the standard undergraduate course in algebra. Following Cornell tradition, I used *Introduction to Higher Algebra* by Maxime Bôcher (formerly of Harvard), an older text with emphasis on the algebra needed for classical higher geometry. Algebraic geometry was a leading specialty at Cornell, led by the veteran professor H. B. Snyder. Graduate students took courses in both analytic and synthetic projective geometry—algebraic surfaces lived on projective space, so that space had to be familiar. I gained an added appreciation of the many uses of algebra through the Bôcher text, and I learned something of algebraic geometry from Robert Walker, who was also a new instructor.

First Teaching

Walker had been a student of Solomon Lefschetz at Princeton; his thesis presented a rigorous form of the famous problem of "resolving" the singularities of a plane algebraic curve, where one made a "rational transformation" to simplify the multiple points on the curve. Despite his help, I was not able to learn enough about algebraic geometry: I only got as far as what Bôcher's book said. However, through Walker I did learn about the sprightly habits of Lefschetz, a Russian immigrant who started as an engineer and became a mathematician after losing both hands in a laboratory accident. With his emphasis on all sorts of varieties of geometry, he vitally influenced algebraic geometry and topology. He was also known for speaking his mind, attested to by the Princeton ditty:

Here's to Lefschetz, Solomon L.,
Irrepressible as hell.
When he's at last beneath the sod,
He'll then begin to heckle God.

Lefschetz did like to heckle his colleagues, as for example the logician Alonzo Church. Church was quite formal, in contrast to Lefschetz's style. He was the first professor of mathematics in the United States whose research focused solely on mathematical logic. He prepared an authoritative bibliography of his subject, and trained many able students, including my friends Stephen Kleene and Barkeley Rosser. He also devised the well-known lambda calculus, codifying the substitution process on functions. Initially, he intended this to be an alternative foundation for mathematics, but Kleene and Rosser later proved such a use inconsistent. A couple of years later, when I returned to Harvard, Lefschetz prepared a treatise on topology that presented a proof of a theorem Hassler Whitney had explained to him. He went through several versions of the proof, sending Whitney and me his revisions every three days; many versions were wrong, but he was willing to accept criticism. Lefschetz was a great advocate of mathematics by word of mouth. One of his decisive contributions to algebraic geometry was to urge people such as Bob Walker and Oscar Zariski to correct the uncertain proofs

Chapter Eight — Cornell and Chicago

presented by the Italians, who, as I mentioned earlier, took an intuitive approach to the subject.

Cornell is on the edge of the small town of Ithaca, New York. White Hall, an older building with a view of Lake Cayuga, housed the mathematics department. The campus is unusually scenic, situated between gorges with waterfalls and streams leading into the lake. The steep walls of the gorges offer prospects of easy exercise in mountaineering. We (the new instructors) all gave it a try; we equipped ourselves with a suitably long rope. In one canyon, I fastened an end of the rope to a tree on the top of the canyon wall while Bob Walker started to climb up the rope from the base. He lost his footing, and for a while swung back and forth like a pendulum; happily, he got out safely. Ithaca was great for outdoor life.

Despite my love for the outdoors, I stayed at Cornell for only a year. In the middle of the year, I was offered an appointment as an instructor at Chicago (I suspect this was an initiative of R. M. Hutchins, who was still president). I decided that Chicago, despite its troubles, provided a deeper mathematics department than Cornell. So, with some regrets, Dorothy and I moved on to Chicago. We found a nice apartment there in the same building as my former high-school teacher, Olive Greensfelder. She was a native of Chicago and had returned home to teach in Gary.

In 1938, our first daughter, Gretchen, was born at the Lying-In Hospital of the university. The doctor's fee was $50 and the hospital bill something like $75, which doesn't sound like much now, but was nevertheless noticeable on an instructor's budget. Her arrival was very welcome, but it added a good bit to Dorothy's duties. However, Dorothy managed to continue to type most of my mathematical papers. We enjoyed watching over Gretchen's development and her subsequent schooling. My mother was very pleased with her, as was her cousin Marjorie, with whom my mother continued to live.

Gretchen soon proved to be an independent, thoughtful child. We have a photograph of her at 3 ½ sitting in a big green arm chair and holding a copy of the left-leaning magazine *New Masses*—I kept that picture secret in later years when I needed clearing to do war research.

First Teaching

```
The MacLane Nucleus

consisting of

Saunders      and      Dorothy
(positive)                        (negative)

has just captured
an orbital electron

Charge              .   .   .   .         negative
Mass                .   .   .   .         3425 grams
Diameter (maximum)  .   .   .   49 cm.
Apparition time     .   .   .   January 5
First Quantum number    .   .   Margaret
Second Quantum number   .       Ferguson
```

Birth announcement, 1938 (note that Gretchen is the German short form of Margaret)

The mathematics department at Chicago had not changed greatly since 1930 when I was a graduate student there. Professor Bliss was still the chairman, and his subject—the calculus of variations—was still dominant. His former student Magnus Hestenes (whom I had known both as a graduate student and at Harvard) returned from UCLA to

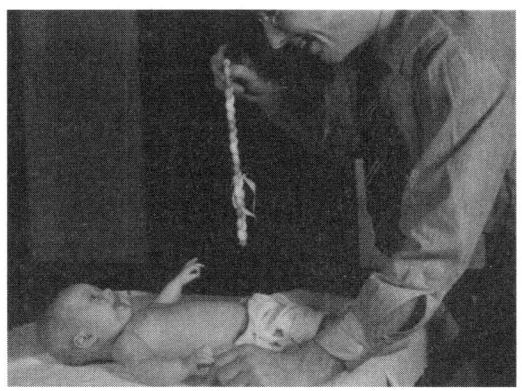

Saunders with his baby daughter Gretchen

Chapter Eight — Cornell and Chicago

Chicago as an assistant professor. Algebra at Chicago had been strengthened with the appointment of Adrian Albert, the outstanding student of Leonard Dickson (Dickson himself was about to retire).

Albert had just written a graduate textbook, *Modern Higher Algebra*; I was assigned to teach the corresponding graduate course, which was inevitably quite different from the course I had taught from Bôcher's text at Cornell. Albert, unlike Bôcher, introduced the Galois theory, though there was a flaw in his proof of the fundamental theorem. In my lecture, I repeated the flaw and challenged students to find it.

Emil Artin, whom I had heard but not understood in Göttingen, was now in the United States at Notre Dame University. He came to Chicago to give a colloquium lecture; he spoke at length and with verve, leaving the impression that he was really thinking out the Galois theory from its first principles right then on the podium. He later recorded his thoughts in an influential booklet, *A Notre Dame Mathematical Lecture, No. 2*.

In addition to the graduate algebra class, my teaching included a section of calculus, taught from the standard, routine text by Granville, Smith, and Longley. It was easy to detect the Yale influence: the explanations of the basic ideas of the calculus were superficial, but the collection of problems for the students to solve was excellent.

The university maintained a downtown branch, where I taught a graduate course on complex variables. My students were mostly professionals with practical interests in the uses of complex variables. What a wonderful subject, with all those amazing consequences of the definition of an analytic function of a complex variable—the first derivate guarantees the existence of higher derivatives. I finally had a chance to expound on what I had learned about this elegant subject in Göttingen from my Ph.D. exam with Professor Herglotz.

The feel of Chicago as a faculty member was different from what it had been as a student. A game of bridge was still the usual social activity, but now Professor Dickson and his wife invited Dorothy and me for dinner and a game (he is said to have played bridge every day at the faculty club). My knowledge of the game was limited but the evening went well. This was to be his last year in office; at the end of the year he left town to return to his native Texas, leaving mathematics behind. I think he could do this knowing that what he accomplished in number theory and algebra—from his history of the theory of numbers to his development of central simple algebras and cyclic division algebras—passed the test of quality.

My own research in algebra had extended from the valuations of fields to the special properties of fields of characteristic p where p is a prime number. These "modular fields" were used in the study of algebraic functions for general (not necessarily complex) coefficients. My research on such fields involved analysis of the lattices of such subfields. At this time, Karl Menger, a Viennese intellectual, had come to Notre Dame after being driven from Vienna by the Nazis. I had first heard of him from Professor Wilson at Yale when he was admiring Menger's new book *Dimension Theory*. His current interest in lattice theory prompted him to organize a conference at Notre Dame on the subject with lectures by many, including Ore, Garrett Birkhoff, and Stone. At the time, Menger, Ore, and Birkhoff were the leading experts on lattices.

I gave a lecture during this conference, in which I presented some of the strange properties of a certain lattice of subfields of a field of characteristic p. I had needed this for my examination of algebraic functions in characteristic p. After I finished, Ore rose to state that there simply could not be a lattice with such properties. But I had such a lattice in hand, and Stone persuaded him to admit that I was right. It has never been quite clear to me why I fell into such repeated disagreements with Ore when he was the person who first introduced me to abstract algebra.

In the spring of 1938, I received an offer to return to Harvard as an assistant professor. Hutchins was still president at Chicago, and he still tended to disagree with physical science faculty members about what and how to teach, but he and the Dean of Physical Sciences both tried to persuade me to stay. The title and the mathematical prospects at Harvard were very attractive, though, so I accepted that offer; however, I continued to respect Hutchins for his deep understanding of the purpose of universities.

Chapter Nine
Surveying Modern Algebra

In the fall of 1938, Garrett Birkhoff and I became assistant professors at Harvard. The mathematics department had, for many years, given a regular undergraduate course in geometry, and Birkhoff pushed for a parallel course in algebra, so Mathematics 6—Algebra—was established. In the initial years, Garrett and I alternated teaching the course. Birkhoff's notes emphasized Boolean algebras, lattices, and groups. My notes (typed by Dorothy) emphasized group theory, Galois theory, and the axiomatic treatment of vector spaces. The latter included an emphasis on linear transformations as the vital background for matrix theory that clearly reflected what I had learned in Göttingen.

Garrett and I combined our presentations to produce a joint book that included both our emphases. For example, the prominent chapter on group theory started with an extensive consideration of examples of groups written to motivate the subsequent, more abstract ideas about groups. For the other topics, we also provided examples first and then theorems. We included a careful treatment of Galois Theory that depended in part on what I had learned from Artin's lecture on the subject in Chicago, and we made a splendid example of abstract methods. Thankfully, Dorothy typed the manuscript for our book.

The publishing firm Macmillan printed the earlier Bôcher text, and we were able to persuade them that an up-to-date book would be useful. As a result, Macmillan published our text *Survey of Modern Algebra* in 1941. It was the first American undergraduate text in algebra that wholeheartedly presented the abstract ideas of Emmy Noether and B. L. van der Waerden. Upon publication, one of my friends told me how it would be received: "Saunders, this text won't fly beyond the Hudson." And so it was, at first. But the abstract view of algebra had many conceptual advantages, and our book made full use of them. Moreover, after the end of World War II, American mathematics became more sophisticated, and for a long period, our survey became the text of choice for undergraduate courses. At one point, there was an abbreviated version entitled *A Brief Survey of Modern Algebra*. Our text is now in its fifth edition, published by A K Peters, Ltd. The various later editions corrected some errors (one never catches them all), added some material, and revised several chapters. However, the revisions were never very extensive; in making them I gathered the clear impression that once a book is written, it is difficult to make substantial changes for later editions, mostly because one cannot easily return to one's original view of the subject.

In terms of writing this book, Garrett and I were lucky to be in the right place at the right time; Harvard had a leading mathematics department during a time when it was becoming clear that undergraduate math majors needed to learn the modern view of algebra. I learned this view from Ore and in Göttingen, and Garrett learned it from Philip Hall while at Cambridge University in England. I had enjoyed teaching modern algebra from various books at several universities, and had learned a great deal about writing from my teacher Olive Greensfelder in high school, and from E. J. Miles at Yale. These advantages worked well for me when writing a more abstract text on algebra and two books expounding on homological algebra and category theory.

Chapter Ten
Algebraic Functions

As already noted, a polynomial equation $f(z, y) = 0$ with complex coefficients determines y as an algebraic function of the complex variable z. The classical theory of these functions, with special attention to their zeros and their poles (the values of z where y "becomes infinite") was a vital part of 19th century mathematics; for example, the solutions could be spread out in several "sheets" over the complex z-plane, giving a so-called "Riemann surface." Felix Klein, in Göttingen, emphasized the visible and intuitive geometric constructions of such surfaces; later, Hermann Weyl elegantly formulated their geometric and conceptual properties in his famous book *The Idea of the Riemann Surfaces*, which gave the first really firm definition of a one-dimensional complex manifold.

In the 1930s, German algebraists generalized this study by replacing the coefficient field C of the complex numbers with an arbitrary field when there could be no usual representation of the resulting algebraic function by a surface, Riemann or otherwise. The situation was especially involved when the coefficient field was of characteristic p (say, the field of integers modulo p). At the Notre Dame conference, I had lectured on certain lattices of such subfields. In Germany, Helmut Hasse and F. K. Schmidt wrote a long paper developing properties of such purely algebraic function fields: I spotted a subtle but substantial error in one of their proofs.

In the meantime, early in 1939, the American Mathematical Society was troubled by some of the new editorial policies used at the

Zentralblatt für Mathematik, published by the well-known German scientific publishing house Springer—there were to be no reviews of German papers by Jewish referees. Because of this, the AMS considered the establishment of a new review journal for mathematical research. Springer sent Heidelberg mathematician F. K. Schmidt to Boston to negotiate with the AMS about review journals; however, this negotiation did not succeed because the AMS was unwilling to consider any special rules that pertained to the race of the author or reviewer of a mathematics paper. Some Americans, notably George Birkhoff, disliked the idea of a new Amercian-managed review journal because they thought it would take valuable research time to write good reviews. Nevertheless, after the conference with Schmidt in Boston, the AMS decided to establish a new international review journal, *Mathematical Reviews*.

My immediate interest in this controversy was Schmidt's presence in Boston—I was able to buttonhole him about the error in the paper he coauthored with Helmut Hasse. We discussed the error, agreed on a repair, and decided to write a joint paper addressing it. This would have been a new collaboration for me, and with a student of Emmy Noether. But our discussion was in the spring of 1939, just a few months before the war broke out. The joint paper never materialized. I subsequently published a correction on my own (in the *Proceedings of the National Academy of Sciences*, 1941). Many years later I visited Heidelberg and talked with Schmidt, but not about this matter—it no longer seemed to be pressing.

The subject of algebraic functions did fascinate me; I also wrote (1940) a lecture notes volume *Algebraic Functions*, based on a course I taught at Harvard. This was before the regular series *Lecture Notes in Mathematics* (later well-managed by Springer), so my students recorded my notes, Dorothy typed them, and Edwards Brothers in Ann Arbor planographed, published, and sold them privately. This was by no means the only early example of the utility of lecture notes, but had a better mechanism been in place, my notes would have, I think, gained the wider publicity they deserved. My notes emphasized the algebraic side rather than the algebraic geometry side, which has subsequently become prominent, as with the proof of Fermat's Last

Chapter Ten — Algebraic Functions

Theorem. They also gave an elegant, then new, proof (due to André Weil) of the basic Riemann-Roch theorem (for the general case), which is a deep theorem connecting the number of poles of an algebraic function with the genus of its Riemann surface.

Chapter Eleven
First Graduate Students

As an assistant professor, I began to direct Ph.D. theses. Here, my first activities turned out to be especially fruitful. At an algebra conference in Chicago I met Irving Kaplansky, who was an undergraduate at the University of Toronto, where he worked with the German group theorist Richard Brauer. Kaplansky entered the first Putnam contest (a prestigious math competition for undergraduates) and won, thereby earning a Putnam Fellowship for graduate study at Harvard. His algebraic interests seemed to fit well with mine, so when he talked to me about research, I suggested that he investigate certain properties of valuations. This led directly to his thesis "Maximal Fields with Valuations" in 1941. It was the start of his long and distinguished career as an algebraist. In view of his earlier successes at Toronto this was no surprise, but it certainly was a very fortunate way for me to begin guiding graduate students.

My next two graduate students were prospective college teachers: Alfred Putnam wrote a thesis about the completion of an integral domain under a given valuation, while his good friend Russell Phelps studied rings with limited associativity. Both came to Harvard from Hamilton College in New York. From the beginning, both were primarily interested in college teaching, and this is what they subsequently did—Putnam in the undergraduate math staff at Chicago, and Phelps at Rutgers and, later, the education division of the National Science Foundation. Their backgrounds at Hamilton interested me because the college is near Utica, where I had gone to grade school, and my father had suggested attending Hamilton.

First Teaching

These first three students illustrate the double role of the Ph.D. thesis in mathematics. Traditionally, the thesis is meant to be an original contribution to knowledge, and can serve well as the first step toward many such contributions, as was the case for Kaplansky. The thesis also qualifies the recipient of the degree to teach mathematics at a college or university, and this teaching can be combined with research work (though it is not required). Both outcomes are socially important, but the path that includes research is perhaps the most demanding. When I started to direct a Ph.D. thesis, I often did not know what the future of the candidate would be. This is, of course, inevitable—the future is indeed inscrutable. Not every thesis (my own, for example) will lead to effective research later, and Ph.D. graduates can also shift fields: W. C. Carter Jr., another student of mine at Harvard, became a computer scientist.

Roger Lyndon was my next Ph.D. candidate at Harvard. I had known him when he was an undergraduate there; his thesis interests covered logic and mathematics. He began as a graduate student in philosophy; when I ran across him he was hopelessly stuck on a thesis problem in logic—a search for axioms for the logic of relations that would resemble the Boolean algebra axioms for the logic of classes. He later found a sophisticated solution for this problem (1956), but for his thesis, I suggested that if he was stuck, he should switch to mathematics and try to do something different. I reported that Eilenberg and I had recently developed a new subject, a cohomology for groups. There were very few examples that had been calculated, and I encouraged him to try some. He did, and found a neat, but subtle method that gave many cohomology groups by successive approximations (taking the cohomology of the cohomology of …). He had a striking result that made a good thesis: for a group G with a normal subgroup N, the method produced the cohomology of G from that of N and of the factor group G/N.

After his thesis was accepted, I chanced to be visiting Paris, where I listened to Jean Leray (then a professor at the College de France) lecture on a way to calculate the cohomology of a space by certain successive approximations, a method he called "spectral sequence." I did not see any connection to Lyndon's work until later, when

Chapter Eleven ⁓ First Graduate Students

Hochschild and Serre used those spectral sequences to calculate the cohomology of a group G with a given normal subgroup. Their method was essentially what Lyndon formulated in his thesis. As a result, I have held that this particular sequence should be called the Lyndon-Hochschild-Serre spectral sequence. These spectral sequences turned out to be a computational device that is much used in algebraic topology and homological algebra, a prospect that was by no means clear to me when I happily accepted Lyndon's thesis. In modern notation, Lyndon had all the terms $E_2^{p,q}$ of that spectral sequence for group extensions, but did not explicitly have the so-called differentials. Lyndon later became a professor at the University of Michigan, and wrote extensively on combinatorial group theory; he was a mathematician of great talent.

While at Harvard, I supervised one more Ph.D. student, B. N. Moyls, on the extensions of a valuation. I also tutored a number of able undergraduate mathematics majors. A good example is Edward Lorenz, who was drafted and sent to meteorology school in preparation to work for the Air Force. When he left the Air Force, he continued to study the new subject meteorology, and became famous for a weather model that showed, through an application of chaos theory, that details of future weather cannot be accurately predicted because of the system's sensitivity to initial conditions. He called this intrinsic unpredictability the "butterfly effect," because the mere flapping of a butterfly's wings somewhere in the world can have unforeseen results in distant locations.

Part Three

Collaborative Research

Chapter Twelve
Crossed Product Algebras and Group Extension

Traditionally, most mathematicians conducted research by working alone, as in the cases of Gauss and Poincaré and many others, eminent or not. Gradually, cases appeared where two people worked together. A notable such case was the joint work in analysis done in England in the 1920s by Hardy and Littlewood. Other such high-level collaborations developed eventually, such as the work in analysis done by Hille and Tamarkin.

In my own case, I came to realize that my research was hampered by a considerable lack of broad knowledge of mathematics. As a graduate student, I had concentrated on logic and foundations so heavily that I had failed to learn enough about lively theories in mathematics proper; overall, I felt a real lack of a mathematics background. Joint work with another active mathematician would, it seemed, help to build up my knowledge. My first experience with joint work, with Virgil Adkisson on point set topology in the 1930s, had brought me into contact with other active topologists. For example, in 1940 I gave a paper on our joint work at the lively conference in topology at the University of Michigan. There, I listened to Wilder, Steenrod, Eilenberg, Whitney, and others, and I realized that I was ready to look for a collaborator or two.

Earlier, sometime in 1937, I had met again with Otto Schilling, a German algebraist I had known briefly at Göttingen. He had been a student of Helmut Hasse, and was now at Johns Hopkins University. As a student of Hasse, Schilling learned much about class field theory, a topic that had intrigued me ever since I failed to understand Artin's lectures in Göttingen. By now, Hasse had codified most of the basic results of the subject (by Hilbert, Artin, and others) in a widely read set of mimeographed notes. I was aware that class field theory gave deep information about a topic of my active interest—the decomposition of rational primes in algebraic number fields with an abelian Galois group over the field of rational numbers. This sophisticated theory made heavy use of "crossed product" algebras, and Schilling and I worked on these algebras together. Because of Schilling's extensive knowledge of class field theory, I managed to learn quite a bit from him.

Crossed product algebras are generalizations of the quaternions, which are linear combinations

$$a + bi + cj + dk$$

with real coefficients a, b, c, and d of four "units" 1, i, j, and k. Addition of quaternions is done by adding corresponding coefficients, while multiplication is given by the following rules for the units

$$i^2 = j^2 = k^2 = -1, ij = k, \text{ and } ji = -k.$$

From these rules, it follows that every nonzero quaternion has a multiplicative inverse; one says that the quaternions form a "division algebra" over the real numbers **R**. Except for **R** and the complex numbers **C**, there are no other division algebras of finite dimension over **R**.

When the complex numbers were admitted as an effective (real) part of mathematics early in the 19th century, the question was left as to whether there were any further such systems of numbers possible. Irish mathematician W. R. Hamilton found the quaternions to be the next such system. Hamilton tried to develop this system for several years, until the missing idea came to him in 1843 in a flash of insight as he walked to Dublin along the Royal Canal. He was so excited by his discovery that he inscribed the multiplication rules for i, j, k on a stone on the Brougham Bridge over the canal. On my one and only

Chapter Twelve ⁓ Crossed Product Algebras and Group Extensions

visit to Dublin, I tried to find this inscription but was not successful. A "pure" quaternion $bi + cj + dk$ can be viewed as a vector in three-dimensional space. For this reason, the quaternions and their associated groups were for a time extensively used in mathematical physics, but the Gibbs vector analysis (after some controversy) largely replaced this use.

The quaternions include the usual complex numbers $u = a + bi$. Since $cj + dk = (c + di)j$ any quaternion can be expressed in terms of the two complex numbers u and v ($v = c + di$) as $q = u + vj$.

Since $ji = -ij$, we also have

$$j(a + bi)j^{-1} = a - bi.$$

Now the function $a + bi \mapsto a - bi$ maps the complex numbers on themselves and preserves both sums and products. In other words, $u \mapsto juj^{-1}$ is an automorphism of the field **C** of complex numbers, which moreover leaves the elements of the subfield **R** of real numbers fixed; this amounts to saying the cyclic group $\{1, i\}$ is the Galois group of **C** over **R**.

The Galois Theory of equations considers generally certain fields K containing a subfield F, just as **C** contains **R**. If K is a so-called "normal" extension of F, then the Galois group G of K over F consists of all the mappings, $x : K \mapsto K$ that are automorphisms (that is, that preserve sums and products and are one-to-one onto) and that leave all the elements of the subfield F fixed.

One can then try to generalize the quaternions by replacing **C** over **R** (complex numbers over the reals) by K over F, as above, and introducing new elements u_x—one for each element x in the Galois group G (u_x is like i, j, k above)—and considering an algebra consisting of elements that are formal sums $\Sigma\, a_x\, u_x$, with each a_x in K, added termwise and with multiplication given by the formulas

$$u_x u_y = f(x, y) u_{xy}, \qquad u_x a = (xa) u_x, \qquad (1)$$

where the indicated factors $f(x, y)$ are scalars in K, one for each pair of group elements x, y in G.

But we still wish this multiplication to be associative. For this, it is enough to require that

$$u_x(u_y u_z) = (u_x u_y)u_z$$

for any three elements x, y, and z of the Galois group G. In this equation, we can calculate both sides by using the product formulas (1). The result (dropping the factor u_{xyz}) is the equation

$$xf(y, z)f(x, yz) = f(xy, z)f(x, y).$$

When the function $f\colon G \times G \to K$ satisfies this equation for all x, y, and z in G, it is said to be a factor set of G over K (today it is also called a two-dimensional co-cycle of G in K). All this indicated that each such factor set will yield an associative algebra of elements $\Sigma a_x u_x$ with multiplication given by (4). This algebra is the crossed product algebra—it is the field K "crossed" with the group G using the factor set f. Schilling and I wrote a long paper on this subject, as well as several shorter papers dealing with formal properties of crossed product algebras that will appear later in our discussion of topology.

It must be noted, however, that there can be a "change of base" for such an algebra, in which each basis element u_x is replaced by $v_x = h(x)u_x$, where h is any function from G to K. One may calculate that this change replaces the factor set $f(x, y)$ by

$$h(x)xh(y)f(x, y)h^{-1}(xy).$$

In particular, the product $h(x)xh(y)\,h^{-1}(xy)$ is itself a factor set, sometimes called a "transformation set."

The American school of linear associative algebra (Leonard E. Dickson, Joseph H. M. Wedderburn, and others) made extensive use of the idea of quaternions. When the cyclic group $\{1, j\}$ of order 2 is replaced by a cyclic group of finite order, one has the "cyclic algebras" studied at Chicago by Dickson. Dickson's student, A. Adrian Albert, once told me that he had generalized Dickson's cyclic algebras by replacing the cyclic group with an arbitrary finite group, thus

Chapter Twelve ~ Crossed Product Algebras and Group Extensions

essentially defining the crossed product algebras. However, on Dickson's advice, he did not publish this idea, so the crossed product algebras were first developed in Germany, notably in a famous paper by Helmut Hasse, Richard Brauer, and Emmy Noether, which rested on earlier German work on group extensions.

In a group extension, each normal subgroup N of a multiplicative group E determines the usual factor group $E/N = G$ and the homomorphism $h: E \to G$ sending each element of e of E to its "coset" eN. The figure is

$$E \xrightarrow{h} G \quad u_x \mapsto x \in G \quad N \to 1 \in G$$

If one chooses for each x in G an element u_x in E which maps by h onto x so that $hu_x = x$, then each product $u_x u_y$ for x and y in G must map onto xy in G and so must have the form

$$u_x u_y = f(x, y) u_{xy}$$

where each $f(x, y)$ is an element of N. Since N is a normal subgroup of G, we must also have $u_x a u_x^{-1}$ in N for each a in N; in other words, the "factor" group G "acts" on the normal subgroup N by

$$xa = u_x a u_x^{-1}$$

for each $x \in G$, $a \in N$. Moreover, the associative law for $u_x u_y u_z$ shows that the function f is indeed a factor set of G with values in the subgroup N. In other words, factor sets can be used to describe the "group" of all given extensions of N by G.

Various algebraists—the Germans Otto Schreier and Reinhold Baer, as well as Marshall Hall and Alan Turing—had studied such group extensions. My own collaboration with Schilling had led me to the study of group extensions generally. Moreover, a little known paper that Schilling and I had written on normal algebraic number fields suggested the possibility that the factor sets $f(x, y)$ perhaps had a higher dimensional generalization. At the time, I did not know that in 1941, Oswald Teichmüller (a fellow student at Göttingen) had found such a three-dimensional generalization to functions $h(x, y, z)$ in a study of algebras. I did know, however, that Noether had studied crossed homomorphisms f of G to N, with

$$xf(y) + f(y) = f(xy) \quad x, y \in N,$$

where the homomorphism f, like the factor set, is "crossed" by way of the operation of x on f.

During this time, I also collaborated on group extensions with Alfred Clifford, an algebraist I had known as an undergraduate at Yale, and it was a series of lectures I gave on group extensions that started my longtime collaboration with Samuel Eilenberg.

Chapter Thirteen
Eilenberg Enters

My vague search for active collaborative research had resounding success in the spring of 1941 when, at the end of my first three-year appointment as an assistant professor at Harvard, I enjoyed a one-semester research leave. I spent some of the time in Chicago collaborating with Otto Schilling, who had been appointed to the department there after my departure. Schilling and I had just completed a long paper on non-abelian class field theory; we were also working on a new and generalized Kummer theory. The class field theory made heavy use of crossed product algebras and their factor sets; some of our calculations suggested that we were missing something beyond factor sets, but we did not know what it might be. Fascinated by these factor sets, I also began to study their uses in the construction of group extensions.

Then, Theophil H. Hildebrandt, the chairman of mathematics at the University of Michigan, invited me to come to Ann Arbor to deliver the annual Ziwet Lectures to the department there. This was an annual series of six lectures, named in honor of a prominent former Michigan mathematician. I could have chosen to lecture on the theory of algebraic functions, as previously described, but I had newer and exciting results on group extensions, so that became my topic.

An extension of a group G by a group H is a group E with normal subgroup G and corresponding factor group $H = E/G$. Such extensions could be described by factor sets of H with values in G. Classically, they were often computed (for example, by Reinhold

Baer, Marshall Hall, and others) by using generators from the group H and the relations between the generators. This amounts to presenting the group H as a factor group $H = F/R$ of the free group F in the generators where R is a (normal) subgroup of all the relations on these generators. In other words, H is the target of a so-called "short exact sequence" of groups and homomorphisms, of the form

$$1 \to R \to F \to H \to 1, \qquad (1)$$

where 1 represents the one-element group. The homomorphism $R \to F$ is the embedding of R, the group of relations, into F, and the map $F \to H$ is the projection of the free group F on the quotient group $H = F/R$. To say that this sequence of homomorphisms is "exact" is simply to say that, in this sequence of homomorphisms, the image of each incoming arrow (homomorphism) is exactly the kernel of the corresponding outgoing arrow. (Here the kernel is the subgroup of all elements mapped into the identity element). Thus, (1) simply states that $H = F/R$.

In my Ziwet lectures I described group extensions calculated by factor sets and by free groups, as above. I emphasized the case of abelian groups, where $\text{Ext}(H,G)$ (in present notation) is the group of abelian group extensions of G by H. In particular, I emphasized the calculation of examples. I especially enjoyed the group H with generators $g_1, g_2, \ldots, g_n, \ldots$ where, for a prime number p,

$$pg_2 = g_1, pg_3 = g_2, \ldots, pg_{n+1} = g_n, \ldots$$

When G is the group of integers, I found that

$$\text{Ext}(H,G) = \text{the } p\text{-adic integers},$$

where a p-adic integer is a sequence $a_0 + a_1 p + a_2 p^2 + \ldots$ with integral coefficients $0 \leq a_i < p$.

In the audience was a young Polish topologist named Samuel Eilenberg, who was then an instructor at Michigan. Eilenberg had studied in Poland until 1939, when his father said, "Sammy, get the

Chapter Thirteen — Eilenberg Enters

Samuel Eilenberg (Photo ca. 1979)

hell out of Poland. War is going to come." Sammy got out and came to Princeton, where Veblen and Lefschetz tried to manage the large flow of incoming mathematical refugees. Sammy's work in topology was well-known in this country, and a position was found at Michigan where there were active topologists.

Sammy listened to my lectures but he had to leave before the last one, so he asked me to give that to him privately, which I did. In this last lecture I calculated the group extension above, and Sammy said to me, "That's just like the Steenrod homology of the p-adic solenoid. Something mysterious is going on here."

He explained that the p-adic solenoid is the limit space formed by taking a torus, wrapping a second solid torus p times around inside the first torus, wrapping a third torus p times around inside the second one, and so on, and forming the common part of all these tori. It chanced that Sammy's colleague Norman Steenrod had found a homology theory of "regular cycles" for "compact metric spaces" that was suited to spaces such as this solenoid. But it was startling to find that its homology could be calculated by group extensions. What did they have to do with regular cycles for homology?

This startling situation kept us up all night, and by morning we had some inkling of what was going on. It was a formula for the connection between the integral singular homology group $H_n(X)$ of a space X and the "singular" cohomology group $H^n(X,G)$ of the same space, with coefficients in an abelian group G. We formulated this result as a short exact sequence

$$0 \to \text{Ext}(H_{n-1}(X), G) \to H^n(X,G) \xrightarrow{\alpha} \text{Hom}(H_n(X), G) \to 0. \quad (2)$$

In other words, the Ext group gives a subgroup of the n-dimensional cohomology group of X and the corresponding quotient group is the group of homomorphisms of $H_n(X)$ into the coefficient group G. The homomorphism α formulated a quite evident influence of homology on cohomology; each "cocycle" of X with coefficients in the abelian group G automatically maps every cycle of X into G. Thus α was quite natural and well-known; what we had found was a description of the kernel of the natural map, and this was where the group Ext of extensions suddenly came in.

It took Sammy and me a while to tease out this connection and to put it into print in a paper "Group Extensions and Homology." Of course, we told Lefschetz about it; he at once asked us to write it up as an appendix (of five pages) to his 1942 book *Topology*. We also published an announcement in the *Proceedings of the National Academy of Sciences* (*PNAS*) and a complete formulation in a paper in the *Annals of Mathematics*. We had found a new intrusion of algebra into topology by the exact sequence (2).

But Sammy was not one to stop with one definitive result. Our first result applied to the usual "singular" homology of a space X. In the simplest case, this is the usual cohomology of a polyhedron. But for more complicated spaces, we needed to get at the homology of a space by taking suitable limits of successive approximations to the space, say by taking finer and finer covers of the space with open sets. This was the well-known Cech homology (say, for a compact metric space). It is all as if one had just two spaces X and Y and a continuous map $h : X \to Y$, and the above short exact sequence for both spaces. In this case, each continuous map h of X into Y carries cycles and homology of X into that of Y. We wrote

$$H_n(h): H_n(X) \to H_n(Y),$$

and for cocycles of Y into X, we wrote

$$H^n(h): H^n(Y) \to H^n(X).$$

This means that h will induce the vertical maps displayed below

$$\begin{array}{ccccccccc} 0 & \to & \mathrm{Ext}(H_{n-1}(Y), G) & \to & H^n(Y, G) & \stackrel{\alpha}{\to} & \mathrm{Hom}(H_n(Y), G) & \to & 0 \\ & & \downarrow & & \downarrow & & \downarrow & & \\ 0 & \to & \mathrm{Ext}(H_{n-1}(X), G) & \to & H^n(X, G) & \stackrel{\alpha}{\to} & \mathrm{Hom}(H_n(X), G) & \to & 0. \end{array}$$

Moreover, the above diagram commutes. By this, we mean that starting on top, going over and then down equals going down and then over. It was fairly easy to show this commutation. We gave it a name; we said that the middle map α was a "natural" homomorphism from cohomology to homology. By natural we meant that it behaved correctly for continuous maps $h: X \to Y$ of spaces. Moreover, we had a good precedent for the choice of the word "natural" here, as I will explain in the next chapter.

With this so-called naturality we were able to show that the exact sequence (2) holds both for the singular homology and for the Cech homology, obtained by suitable limits. Since this sequence essentially gives a way to compute the cohomology H^n with coefficients G from the integral homology groups H_n and H_{n-1}, it was called a universal coefficient theorem because it calculates cohomology for any coefficient group G from the integral homology using the new object "Ext" of group extensions.

Eilenberg and I ultimately wrote 15 joint papers that had substantial effects—it was fortunate for us. For my semester leave from Harvard, I had actually applied to the Institute for Advanced Study to work there. My application was turned down, but this turned out to be for the best. Had it been accepted, I might have missed working with Sammy.

Chapter Fourteen
Naturality

The universal coefficient theorem starts with a homomorphism α from the cohomology $H^n(X,G)$ of a space X with coefficients in a group G:

$$\alpha: H^n(X,G) \to \mathrm{Hom}(H_n(X),G).$$

Moreover, this homomorphism behaves well for continuous maps $f: X \to Y$ of spaces. In the last chapter, I said that *we* gave this simple commutativity behavior the name *natural*. This is not really accurate. We didn't come up with this term—we just followed the informal terminology then used for such matters.

A known example was that of the dual V^* of a vector space V over a field F. The definition of the dual space is just

$$V^* = \mathrm{Hom}(V,F).$$

In other words, the dual space V^* consists of all the linear transformations $L: V \to F$. Then, given any linear transformation $f: W \to V$ from another vector space W over the same field F, one has the evident composite $Lf: W \to F$. In other words f gives a map

$$f^*: V^* \to W^*, \tag{1}$$

sending each element L in V^* into Lf in W^*. This means that the process of forming the dual vector space applies not just to vector spaces, but also to linear transformations between such spaces. Thus, the dual operation applies both to spaces and to transformations of spaces.

There is also a well-known map ϑ sending each space V to its double dual V^{**}. Indeed, define $\vartheta : V \to V^{**}$ by setting

$$(\vartheta v)L = L(v), \text{ for } v \in V, L \in V^*.$$

One easily sees that this ϑ is linear both in v and in L, and so gives a linear map

$$\vartheta : V \to V^{**}, \, v \mapsto \vartheta v.$$

If V is of finite dimension, ϑ is actually an isomorphism. One already knows that dim V = dim V^*, therefore dim V = dim V^{**}.

Hence, one can find many such isomorphisms $\vartheta : V \to V^{**}$ by picking any basis for V and sending it to any basis for V^{**}. But the special isomorphism ϑ above is defined by first principles, independently of the choice of a basis. For this reason, mathematicians generally said that this ϑ was "natural."

Sammy and I observed that this naturality also means that when each V itself is varied by a linear map $f : W \to V$, there is a dual map $f^* : V^* \to W^*$ as in (1), and hence a double dual $f^{**} : W^{**} \to V^{**}$, which makes the following diagram commute

$$\begin{array}{ccc} W & \xrightarrow{\vartheta} & W^{**} \\ \downarrow f & & \downarrow f^{**} \\ V & \xrightarrow{\vartheta} & V^{**} \end{array} \qquad (2)$$

We concluded that the naturality of ϑ (for W and V) was connected to the fact that ϑ commutes, as in (2), with the maps f and f^{**} induced by any f. This is part of a larger diagram for three vector spaces, as follows:

Chapter Fourteen ⁓ Naturality

$$\begin{array}{ccc} W & \xrightarrow{\vartheta} & W^{**} \\ \downarrow f & & \downarrow f^{**} \\ V & \xrightarrow{\vartheta} & V^{**} \\ \downarrow g & & \downarrow g^{**} \\ U & \xrightarrow{\vartheta} & U^{**} \end{array}.$$

Here the category on the left of vector spaces is mapped to itself by the double dual functor, and ϑ is a natural isomorphism of the identity to the double dual functor.

This example, and others, leads directly to the general definition of category, functor, and natural transformation.

A *category* is a collection of objects $A, B, C, \ldots$ and morphisms $f, g, \ldots$ Such that

$$A \xrightarrow{f} B, \qquad B \xrightarrow{g} C \qquad (3)$$

Each morphism f has a domain A and a codomain B. Whenever g has domain B = codomain f the compound morphism gf

$$gf: A \longrightarrow C$$

is defined. This composition of morphisms is associative; that is, when $h: C \to D$, then

$$h(gf) = (hg)f: A \to D.$$

Moreover, for each object B there is an identity morphism $1_B: B \to B$ with $1_B f = f$ and $g 1_B = g$.

This describes categories. Next come *functors*, the maps of categories.

If $\mathcal{C}$ and $\mathcal{E}$ are categories, a *covariant functor* $F:\mathcal{C} \to \mathcal{E}$ is a pair of functions F_0 and F_1 sending objects and morphisms of $\mathcal{C}$ to objects and morphisms of $\mathcal{E}$ in such a way as to preserve domain, codomain, identities, and composites. Thus, given (3) in a category $\mathcal{C}$, one has

$$F_1(f):F_0(A)\to F_0(B), \quad F_1(g): F_0(B)\to F_0(C),$$

as well as

$$F_1(gf)=F_1(g)F_1(f), \qquad F_1(1_B) = 1$$

On the other hand, a *contravariant functor* $H:\mathcal{C}\to\mathcal{E}$ is a pair of funtions H_0 and H_1 again sending objects and morphisms of $\mathcal{C}$ to objects and morphisms of $\mathcal{E}$ so as to preserve identities while reversing domain, codomain, and composites, as in the equations

$$H_1(f) : H_0(B)\to H_0(A), \quad H_1(g) = H_0(C)\to H_0(B),$$

and

$$H_1(gf) = H_1(f)\, H_1(g), \quad H_1(1_B) = 1.$$

One may also define the opposite of a category $\mathcal{C}$ to be the category $\mathcal{C}^{op}$ with the same objects and arrows, but with composition and the direction of all arrows reversed.

In $\mathcal{C}$,
$$A \xrightarrow{f} B, \quad B \xrightarrow{g} C, \quad gf: A \to C.$$

In $\mathcal{C}^{op}$,
$$B \xrightarrow{f^{op}} A \quad C \xrightarrow{g^{op}} B \quad f^{op}g^{op}: C \to A.$$

Thus, a contravariant functor on $\mathcal{C}$ is the same thing as a covariant functor on $\mathcal{C}^{op}$.

Chapter Fourteen ⇒ Naturality

One may also define the product of two categories $\mathcal{C}$ and $\mathcal{C}'$. The objects are pairs of objects A, A' and B, B' with A and B in $\mathcal{C}$ and A' and B' in $\mathcal{C}'$, the arrows are pairs of arrows f, f', with $f: A \to B$ in $\mathcal{C}$, and $f': A' \to B'$ in $\mathcal{C}'$. A functor on a product category $\mathcal{C} \times \mathcal{C}'$ is also called a *bifunctor*.

These definitions are simple but very general—so much so that Eilenberg and I did not immediately publish them in all their glowing generality. Instead, our first paper on categories was a note in the *Proceedings of the National Academy of Science* describing just functors on the category of groups or that of abelian groups. The direct product $G \times H$ of two groups is a bifunctor of groups to groups, as is the free product of two groups. For the category of abelian groups the usual direct product $G \times H$ as well as the tensor product, $G \otimes H$, are both bifunctors on abelian groups to abelian groups. If G and H are both abelian groups, the set $\text{Hom}(G,H)$ of all homomorphisms $G \to H$ is also an abelian group, and in this case there is a natural isomorphism

$$\tau: \text{Hom}(G,\text{Hom}(H,K)) \cong \text{Hom}(G \times H, K) \qquad (4)$$

of trifunctors, covariant in K and contravariant in the groups G and H. Indeed, this isomorphism $\tau = \tau(G,H,K)$ assigns to each homomorphism $\phi: G \to \text{Hom}(H,K)$ the homomorphism $G \times H \to K$ defined for any generator $g \times h$ of $G \times H$ as

$$([\tau(G,H,K)]\,\phi)\,(g \times h) = [\phi(g)](h),$$

an element of K. The isomorphism (4) above states that the functor $— \phi H$ is a left adjoint to the functor $\text{Hom}(H,—)$. It was an early example of the notion of adjoint functors that later became so important. But the importance could not have been seen at that early date. At that point in time, the theory of categories seemed to be an exceedingly general topic that covered many examples. One can consider the category of topological spaces, where morphisms are the continuous maps of such spaces, or the category of rings where morphisms are the homomorphisms of rings, or say the category of sets, where morphisms are the functions—and so on.

Thus, it was that my initial collaboration with Eilenberg produced on the one hand a needed explicit formula (for the solenoid), and on the other hand, the very general notions of category and functor.

Part Four

The War Years

Chapter Fifteen
Much More Applied Math

With the onset of the war in Europe in 1939, assertions arose that there was not enough study of applied mathematics at universities in the United States. This was both a general observation and a call to do something better in light of the prospective need for the uses of mathematics in war research. In response to these observations and criticisms, there were efforts to improve, and I will mention a few here.

At Brown University, which to this day has a strong division of applied mathematics, Dean R. G. D. Richardson led the reform. For many years, he had been the secretary of the AMS, and in this capacity, had acquired a wide circle of acquaintances. In the late thirties, when Brown appointed new faculty members, they included William Prager, who had taught in Berlin and fled from the Nazis to Turkey. He happily accepted the invitation to come to Providence, where he taught courses in his specialty, plasticity theory. The university set up a program to retrain younger American mathematicians in applied mathematics. This program attracted many who thought of taking this opportunity to brush up on their skills in applications—I was one of them. I traveled from Cambridge to Providence to learn about the mathematics of plasticity and related subjects, including some background work in elasticity. I found the program interesting, but not as exciting as algebraic topology. I made no explicit use of the knowledge I gained from this program in my subsequent war research, but this did not trouble me.

The War Years

NYU also experienced changes: Richard Courant had been a professor and administrative head of the Mathematical Institute of Göttingen during my studies there, and had worked extensively with differential equations, conformal mapping, and Plateau's problem. Like many, he came to the U.S. to escape Hitler, although somewhat later than his colleague Hermann Weyl. Apparently it was Weyl who found a position for Courant at NYU. Before Courant's arrival, there was little active mathematical research at NYU, but he set out to change that. He brought in several of his former students from Göttingen, including Kurt Friedrichs and Fritz John, and he encouraged American mathematicians that were at hand, for example, James J. Stoker. He quickly noted that many promising younger mathematicians could be found in New York City, such as Peter Lax. With various collaborators, Courant completed the long intended "Courant-Hilbert" book, *Methods of Mathematical Physics*, which became an influential textbook on partial differential equations and related matters.

Within one year of his arrival, he had created a lively mathematics institute, later to be called the Courant Institute for Mathematical Sciences. From time to time, Courant and I met and renewed our Göttingen acquaintance. Back then he had seemed to me to treat all of mathematics impartially; now he was an ardent proponent of applied mathematics, and it worked. For example, I recall a wartime paper by NYU mathematicians titled "Water Waves on Sloping Beaches." The subject was not accidental; it was clearly connected to the American Navy's island hopping campaign in the Pacific. All told, Courant's story was a rousing success; he set up applied mathematics at NYU and went on to do applied mathematics research useful for war and peace.

After the war, Princeton encouraged vital developments in mathematical statistics. Samuel Wilks had been trained in statistics at the University of Iowa and then joined the Princeton faculty, where he encouraged many subsequently influential mathematical statisticians. For example, John Tukey's influential Ph.D. thesis (1940) "Convergence and Uniformity in Topology" involved a sophisticated study of the use of the Axiom of Choice in topology.

Chapter Fifteen ⁓ Much More Applied Math

After this abstract start, he soon became engaged with Wilks in war research projects. By 1946, he was publishing papers in statistics, and became famous for his contribution to the Fast Fourier Transform. This and other cases suggest that talent for mathematics can switch from pure to applied.

These examples indicate that from the 1940s on, the American mathematical community rapidly took on applied mathematics and pressed ahead. In addition to the activities at Brown, NYU, and Princeton, there were numerous other movements toward applied mathematics. But was applied mathematics neglected in the United States before the World War II? An examination of several major departments can add insight to this question.

In the 19th century, celestial mechanics was a central part of applied mathematics in the U.S. George William Hill (1838–1914) of the Nautical Almanac Office and Simon Newcomb (1835–1909) worked on the fundamental constants of astronomy and taught at Johns Hopkins University.

My experience at Yale illustrates the early, but diminishing, importance of celestial mechanics. My teacher E. W. Brown was a member of the Yale faculty for many years; he represented applied mathematics, but not in a way that stimulated students to follow in his footsteps. Physicist J. W. Gibbs had students with mathematical leanings, but none remained at Yale in mathematics. Edwin B. Wilson, for example, collaborated with Gibbs on their influential book on vector spaces; he was a mathematics instructor for a time, but was tempted away to a better position at MIT. When I was an undergraduate at Yale, there was no excitement present in the applications, not even in theoretical physics—celestial mechanics no longer seemed to be a hot subject.

At Harvard, E. V. Huntington had a chair intended for applied mathematics. He taught statistics, but was probably more interested in finding new axioms for Boolean algebras. Around 1930, the Harvard administration pushed for more applied mathematics, which resulted in the appointment of J. H. van Vleck, who joined the mathematics department but kept a primary interest in physics, the subject in which he won his Nobel Prize. In general, the connection between mathematics and mathematical physics became weaker with time.

At Princeton there was perhaps more applied mathematics; take, for example, the active mathematical physicist H. P. Robertson. He was not immune to the customary Princeton ditties:

> Here's to Robertson, Howard Percy
> On his soul there'll be no mercy
> Light of finger and deft of toe
> His brow is high, but his mind is low.

In the fine days of E. H. Moore in the 1890s, Chicago had been active in applications, for example, with the eminent applied mathematician F. R. Moulton. But Moulton retired, and applied math had become weaker, before I came as a graduate student in 1930.

From 1910–1920, the University of Wisconsin had an active group in mathematical physics—Max Mason and Warren Weaver wrote a well-regarded text on the electromagnetic field. Mason left to become president at Chicago; Weaver became vice president of the Rockefeller Foundation, and as mentioned earlier, Harvard recruited J. H. van Vleck.

These examples taken together suggest that at number of leading American universities, effective representation of applied mathematics was often lost or minimized. I suggest it was often for lack of enough new ideas or new directions for applied research—the exciting developments in mathematics come from new ideas, not just from tradition.

Chapter Sixteen
Cynthia Enters

Our second daughter, Cynthia Robbins, was born in May 1941 while I was on a semester leave from Harvard. At the time, I had finished the Ziwet lectures in Ann Arbor, and was studying in New Haven. Thus, Cynthia arrived in the New Haven Hospital. Dorothy had real trouble—she suffered a severe hemorrhage after the birth. Fortunately, the nurse in attendance caught it in time to order treatment, but it was a severe shock for Dorothy and me. After the birth, Dorothy and Cynthia moved to Norwalk, Connecticut, to be with my mother and Marjorie; our older daughter, Gretchen, was already there. It was good to have their help with the new baby.

Norwalk, with my mother and Marjorie Harrington, was a happy spot for our children. Marjorie lived in the home at 13 Eversley Avenue that she had inherited; it was a commodious and comfortable house with a large garden in the backyard. Marjorie had once been active as a teacher, and took great pleasure in the presence of our daughters. She captured their youth in storybooks, drawings, and verse.

One of her verses referred to the girls' rocking horses (Dapple Gray was a traditional rocking horse, and Dom Pedro and Snow Drop were two horses connected by a baby seat):

> Hurray, Hurray! We're off and away!
> And because I am bigger I ride Dapple Gray
>   For he is a terribly spirited horse
> But Cynthia's little, and so she, of course,

> Must ride with Dom Pedro and Snow Drop so slow
> Now out of the way folks, and see
> How we GO!

Another poem in the book from Gretchen's point of view:

> When I was little, I crept alone
>   For Cynthia wasn't there.
> And now in spite of the fact I am grown,
>   I go creeping everywhere.
>
> If it wasn't for me, how could she learn?
>   For creeping's a tricky art.
> It is always a question which way to turn,
>   And a question how to start.
>
> Oh, I teach Cynthia carefully.
>   But who in the world was it
>   That taught me?

My mother and Marjorie taught our daughters many things.

As for their care, Dorothy did most of the chores—I didn't feed them or change them, although I could. I like to think that this was typical of fathers during that time; most evenings, I worked in the study doing mathematics. But I enjoyed playing with our daughters very much. A neighbor once told a friend of ours that she liked to watch out her back window when I came home, when I would take the girls around on my back and do all sorts of stunts with them.

Chapter Seventeen
War Research: Roll, Pitch, and Yaw

In the spring of 1943, I began to really learn about mathematical war research: the National Defense Research Council (NDRC) was reorganized, and for the first time included a panel on applied mathematics, headed by Warren Weaver. This panel was designed to oversee various groups working under contracts at several universities. Weaver, who probably knew me from the AMS, approached me with the proposal that I come to New York to work in the just-established Applied Mathematics Group, Columbia (AMG-C). His suggestion involved a specific problem about gases diffusing through a complex sequence of pipes, which sounded difficult but fascinating. I agreed to come, took leave from Harvard, and so landed at Columbia, though once I arrived I never again heard of this particular problem. Much later, with the publication of the Smyth report on atomic energy, I concluded that the problem had to do with the gaseous diffusion process for separating isotopes of uranium.

At AMG-C, the actual problems were chiefly those that came to us from the Air Force. At that time, U.S. bombers were flying over Germany, where they were attacked by German fighter planes. The bombers were defended by machine gunners, so the essential problem involved how to aim the machine guns at the fighter planes. As background, we needed to study motion of airplanes; these angular motions involved essentially three angles of roll, pitch, and yaw. Those in turn involved spherical trigonometry, a standard subject none of us had studied. However, we did learn it in one active evening—new mathematics can indeed be acquired.

How should a machine gunner aim at an attacking fighter plane? The fighter is approaching the bomber, but the bomber is moving forward at the same time. The resulting rule is that the gunner should aim toward the tail, which is opposite from the rule for hunting ducks, where the rule for a stationary hunter is to aim ahead of the ducks. A major part of our problem was properly training machine gunners to aim toward the tail.

In addition, certain lead-computing sights were in use, which were equipped with gyroscopes; as the gunner tracks the oncoming plane, the gyroscope records the angular motion of the line of sight and, from this, calculates an amount and direction of lead. Engineers had designed the Mark 18 sight to do this task. Earlier, Warren Weaver had made some analyses of how well this actually worked out; he gave the analyses over to me for completion, and it turned out that in many cases the sight did not get the lead right. I gathered the impression that the engineering design had not been subjected to sufficient mathematical analysis—we had to provide this after the fact, which meant that our contribution to the war effort came too late.

The Air Force also needed help in the field with the training of the actual gunners. At AMG-C we enlisted about 18 willing mathematicians who brought training instructions with them overseas to the bomber bases where they trained the gunners.

Many other elementary mathematical problems arose—AMG-C expanded, and when I became the director in 1944, I moved my family to New York. I took the occasion to hire a number of mathematicians well-known to me, including Samuel Eilenberg, Paul Smith, and Hassler Whitney. Whitney turned out to be especially effective, not in applying his expert knowledge of algebraic topology, but in talking to officers at various bases and discovering what the real problems at issue might be. Many old friends and acquaintances worked at AMG-C: Walter Leighton, with whom I had studied Banach spaces at Harvard; Magnus Hestenes from Chicago; Arthur Sard, who had studied with Marston Morse at Harvard; Irving

Chapter Seventeen ~ War Reasearch: Roll, Pitch, and Yaw

Kaplansky, my former student; and Daniel Lewis, who had been a fellow instructor at Cornell. At AMG-C and in other similar groups, it turned out that general mathematical talent was effective; it could be applied successfully to the practical engineering problems that arose.

A typical explicit problem was the description of pursuit curves. The simplest example is two-dimensional: A point P in a plane moves along a straight line at constant velocity; another point Q moves in the same plane at a different constant velocity, but so as to be always aimed at the present position of P. What curve (a pursuit curve) does Q follow? This problem had been variously studied, but our technical aide, Dr. E. W. Paxson, found that the real problem was not planar but three-dimensional. Dr. Leo Cohen, a topologist converted to work at AMG-C, in cooperation with John Tukey from Princeton, found the correct differential equations for this problem. Explicit solutions were what mattered, and they were computed from the equations by dozens of young women at desktop calculators.

This is just one example of the varied mathematical problems that came to our attention. Repeatedly, we needed to consult Air Force officers to understand what the real problems were. We also heard of statistical problems that came up to the Applied Statistics group at Columbia, which was in the same building as our group and directed nominally by Harold Hotelling, but in actuality by Allen Wallis. From them and from others we learned about some of the problems of operations research, a field of study developed in connection with the war against the submarines.

Inevitably, we all had official security clearances so that we could see the various classified reports as they circulated, which came under various levels of security such as Restricted, Confidential, Secret, or Top Secret. There was even a classification Burn After Reading, and it was rumored that some extremely classified documents were classified as Burn Before Reading. But all these levels of security didn't mean that they properly classified everything—there were many tables of trigonometric functions that were classified. Whatever the case, we kept such items under lock and key, all while noticing that even nonsense can be considered confidential. We did learn that the real problems confronting the Air Force were not necessarily correctly described in the classified literature—direct contact with the Air Force provided much better information. In theory, the members of the Applied Mathematics Panel of the NDRC formally instructed us what to study; in practice, we often had to find this out directly from the Air Force. We carried out more than one study before it was officially authorized.

Part Five

Eilenberg and Mac Lane

Chapter Eighteen
Cautiously Publishing Category Theory

At Columbia, Sammy and I frequently spent the day working on Airborne Fire Control and then, in the evening, moved to his apartment to consider ordinary mathematical problems. Our collaboration had started with two different objectives. On the one hand, we had answered an explicit and concrete problem: How can you compute the cohomology of the p-adic solenoid? This is the problem that led to our universal coefficient theorem. On the other hand, this theorem had suggested the very general and abstract notions of category and functor. These notions were so general that they hardly seemed to be real mathematics—would our mathematical colleagues accept them?

While we were at Columbia we prepared a careful, systematic exposition of these ideas in a manuscript titled, "The General Theory of Natural Equivalences," which we submitted to the *Transactions of the American Mathematical Society*. Paul Smith was the current editor who knew Sammy well from common interests in topology, so Sammy talked to him about possible referees for our paper on categories. Sammy suggested his young friend George Mackey, who, as a former student of Marshall Stone, was definitely interested in abstract ideas. I do not recall what was in his referee report, but the paper was accepted by the *Transactions*. As a result, the general theory of categories was published in a timely way, proudly discovering very general conceptual aspects of mathematics. At the time, we sometimes called our subject "general abstract nonsense." We didn't really mean the nonsense part, and we were proud of its generality. At the time, Sammy decreed that there was no need for further publication on categories—with functors and natural transformations, it was all already there.

Algebraic topologists had long realized that this subject worked by making algebraic pictures or images of geometric situations. Brouwer's famous theorem on degrees specified that two continuous maps of a sphere onto a sphere could be deformed into each other if and only if they had the same degree; in this case of the degree, a single number (an algebraic object) answers the geometric problem at hand. More generally, the homology and cohomology groups of a space describe properties of that space and vary algebraically with continuous maps of spaces. Categories provide an explicit formulation of this idea as follows: The nth homology group of a space is a functor on the category of spaces to the category of groups. Moreover, cohomology groups are also functors—contravariant ones. This simple idea, perhaps formulated vaguely without the use of categories, was clearly present in algebraic topology in 1940, even before there was any formal theory of categories. The presence of the category made it possible to formulate the idea explicity, as seen by Sammy and Steenrod in their 1952 book *Foundations of Algebraic Topology*, where they formulated axioms on the relative homology $H_n(X, A)$ of a space X modulo a subspace A. H_n is described as a suitably axiomatized functor on pairs of spaces to groups. This book also effectively organized the various different homology theories then in use. It is notable that their formulation made essential use of $H_n(X, A)$, because this concept had been used extensively in Lefschetz's books on topology. The axiomatization made its importance clear, as expressed by a boundary homomorphism: $H_n(X, A) \to H_{n-1}(A)$.

The use in topology was the first effective application of category theory. Many other applications developed later, often by Sammy's students. The general idea of category theory was indeed useful; we were fortunate to have found it and published it at an early time.

Chapter Nineteen
The Cohomology of Groups

A topological space X has both homology and homotopy groups as invariants. The homology groups arose from the study of the Betti numbers of a space. Another invariant, the fundamental group $\pi_1(X)$ of a space X was also well known. It depends on a choice of a "base point" $x_0 \in X$. Its elements are loops; that is, they are continuous maps $f: I \rightarrow X$ of an interval I onto the space X, where both endpoints 0 and 1 of the interval are mapped into the base point. Two paths f and f' count as equal when the first can be continuously deformed into the second. If $g: I \rightarrow X$ is a second such path, the product fg is defined by following first f, then g; using the definition of equality, one sees that the product is associative, but not necessarily commutative. For example, if the space is a plane with the points a and b deleted, the path from the base point looping once around a does not commute with the path running once around b. The fundamental group $\pi_1(X)$ of a space X is closely related to its one-dimensional homology group $H_1(X, Z)$, with integral coefficients Z. This homology group is abelian and is, in fact, isomorphic to the abelianization of $\pi_1(X)$; that is, the group $\pi_1(X)$ divided by the subgroup of all products of commutators (the commutator of paths f and g is the path $fgf^{-1}g^{-1}$). In this way, the fundamental group determines the one-dimensional homology.

But what happens in higher dimensions? In 1932, at the International Congress of Mathematicians, the Czech topologist, Eduardo Čech introduced higher homotopy groups $\pi_n(X)$. In his definition for $n = 2$, an element of $\pi_2(X)$ is a continuous mapping of a square $I \times I$ into the space X such that the boundary of the square is mapped into the base point x_0. The product of two such elements f and $g : I^2 \to X$ of π_2 is formed by pasting two squares together along one edge and mapping the first square by f, the second by g (and all the edges going into the base point), as in the figure

As before, two elements f and $f' : I^2 \to X$ count as the same element of π_2 when the first map f can be continuously deformed into the second one, holding the boundary points fixed. This definition has the effect that the product fg is equal to the product gf, as indicated by the following deformation:

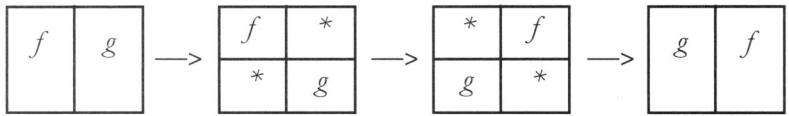

With this definition it follows that $fg = gf$. In other words, the so defined second homotopy group $\pi_2(X)$ is abelian! This result is very much unlike the first homotopy group $\pi_1(X)$, which, in general, is *not* abelian. When this startling circumstance appeared during the discussion at the conference, Čech withdrew his definition of $\pi_n(X)$—he thought he must be mistaken. In particular, he studied spaces X in which only one homotopy group $\pi_n(X)$ was nontrivial. He proved that in such a space, the homotopy group π_n determines all the homology groups of the space, but he did not try to make this determination explicit, which would have been difficult at that time.

In 1942, Heinz Hopf, in a remarkable paper, examined this situation and showed for a polyhedron P that if $\pi_2(P) = \ldots = \pi_{n-1}(P) = 0$, then $\pi_1(P)$ determines part of the integral n-dimensional homology of P. More exactly, let $S_n(P, Z)$ be that subgroup of $H_n(P, Z)$ that

Chapter Nineteen ⁓ Cohomology of Groups

consists of cycles coming from mappings of the n-sphere into P. Then $\pi_1(P)$ determines the quotient group $H_n(P, Z)/ S_n(P, Z)$. Moreover, Hopf gave an explicit, algebraic formula in π_1 for this quotient group. In particular, if all the higher homotopy groups of P are zero, this means that the fundamental group $\pi_1(P)$ determines all the homology and cohomology groups of P. Astounding!

Eilenberg suggested to me that we might use our techniques to find a better formula for this determination. To do this, we used Eilenberg's recent systematic formulation of singular homology, which used the singular complex of a space X, often written as $S(X)$. There, an n-dimensional simplex was a continuous map $\Delta_n \to X$ of an n-simplex Δ_n into the space X, with boundary given by the sum of the usual faces. For a space X with all higher homotopy vanishing, this complex could be described directly in terms of $\pi_1(X)$. The cohomology of this complex with coefficients in an abelian group G depended only on the group π_1, so is now called the cohomology $H^n(\pi_1, G)$. In other words, where $\pi_n(X) = 0$ for $n > 1$, then

$$H^n(X, G) = H^n(\pi_1, G).$$

Thus, the cohomology of the space depends only on the fundamental group π_1! Moreover, the formulas for this group cohomology $H^n(\pi_1, -)$ were explicit—a two-dimensional cycle was exactly a function $f: \pi_1 \times \pi_1 \to G$ with zero coboundary δf, for x, y, z in π_1:

$$\delta f(x, y, z) = x f(y, z) - f(xy, z) + f(x, yz) - f(x, y).$$

In other words, it was exactly a factor set for π_1 in the sense described above. The formulas in higher dimensions were analogues. In this way, following the initiative of Hopf, we were able to see that algebraic objects like groups indeed had cohomology, and that this cohomology of groups was a direct generalization of the theory of group extensions.

In retrospect, the discovery of group cohomology in this sense was an inevitable application of topology to algebra. After our initial discovery, Eckmann, Freudenthal, and Hopf all found equivalent formulas. This all happened during the war, so the news didn't get around very promptly; however, it did initiate the idea of applying concepts of homology to algebraic systems: groups, rings, algebras, and others. This development was one of the main aspects of a new subject called homological algebra, as later organized in the Cartan-Eilenberg book of that title, and in my related book *Homology*.

Chapter Twenty
Eilenberg-Mac Lane Spaces

The discovery of group cohomology gave the cohomology of a space X where the fundamental group π_1 is present, but all higher homology groups vanish. This result raised an evident question: What happens with a space X with only one nonvanishing homology group $\pi_n(X)$ for $n > 1$?

It turns out that in this case also, all the homology and cohomology of X is exactly determined by this one group $\pi_n(X)$, but the formulas for this determination become much more cumbersome. To get at them, one needed to study the singular complex $S(X)$ of such a space. This complex, call it $K(\pi, n)$, could be explicitly given in terms of Eilenberg's singular homology theory, which makes extensive use of the operators face and degeneracy. Each n-dimensional simplex σ has an ith face $f_i\sigma$ and an ith degeneracy $s_i\sigma$ for $0 \leq i \leq n$; the same systematic equations relating these operations define what is indeed a simplicial object. We could write down these equations for one of those spaces $K(\pi, n)$ and then write down the very involved related equations for $K(\pi, n + 1)$. We called this construction W from n to $n + 1$ and wrote $W(K(\pi, n)) = K(\pi, n + 1)$. It seemed hopelessly complicated.

After a long and arduous computation—the worst I have ever completed—we did find an efficient way to describe the passage from $K(\pi, n)$ to $K(\pi, n + 1)$. This passage involves a product in $K(\pi, n)$ where the product of two simplices $[x_1,...,x_p]$ and $[y_1,...,y_q]$ was obtained by using a signed sum of all the ways of shuffling the ys through the xs. Finally, using this shuffle product, we found an elegant construction to get $K(\pi, n + 1)$ from $K(\pi, n)$. We baptized this construction the bar construction because the symbol we used to separate adjacent cells was a bar. With this, we finally found a satisfying way of describing those spaces $K(\pi, n)$ with only one homotopy group π in dimension n, which are now called Eilenberg-Mac Lane spaces. I first described them in a public lecture at the Harvard Mathematics Colloquium. Hurewicz was in the audience, and did not, at the time, wish to believe that such spaces could really matter, although one of his earlier studies had stated that in such a space the group π_n (the only nonzero homotopy group) would determine all the homology and cohomology groups of the space. What we had found was an array of specific formulas describing exactly this determination, and later work by Henri Cartan and others made many improvements on it.

Part Six

Harvard Years

Chapter Twenty One
Professor at Harvard

My years at Harvard, 1938–1947, were a fine time to be in a first-class department at a leading university. I eventually rose in rank from assistant to full professor, with an interruption for war research. We held department meetings at the home of the chairman, who, at the time, was Joseph Walsh. During meetings we discussed department policy and potential new appointments to the department, although new appointments were rare.

I was once a member of a subcommittee to look into junior appointments; we chose George Mackey, a recent Ph.D. student of Marshall Stone. We also reappointed Lars Ahlfors, who had left the department to return to Finland during the war. However, we did not make any senior appointments from outside the department; we could have considered the possibility of appointing André Weil, an outstanding French mathematician who had become a refugee during the war, but we never did. I was quite aware of his abilities—at one point, I persuaded the AMS to invite him to give an hour address at one of its New York meetings, and he gave a good address on some of his important discoveries in algebraic geometry. Eventually, he did come to an American university, going first to Lehigh, and then, after an appointment at the University of São Paolo, Brazil, he went to the University of Chicago.

We did appoint new Benjamin Peirce instructors; for example, Leonidas Alaoglu, a Chicago Ph.D. who I had formerly known there. As usual, he was put to teaching Osgood's text for calculus, which

Cynthia and Gretchen in the backyard of 7 Avon Street, April 1947

included a chapter discussing infinitesimals. "Foggy" Osgood thought that the formal discussion of limits was a bit much for Harvard freshmen, so he used the older notion of infinitesimals. Since then, many senior mathematicians have been doubtful about the concept; apparently, Alaoglu was too—the following was reported from a session of his class: "Gentlemen, we now come to Chapter VII on infinitesimals. Please take pages 123-150 between thumb and finger of the right hand, and tear the pages from the book." And it was done. I do not believe he was subject to any official sanction for this act—Harvard customs were firm, but deviations were not necessarily publicly punished.

At first, Dorothy and I lived in an undistinguished, though big, apartment; it was U-shaped, as had once been the case for Marshall Stone. But we then found a two-family house, divided vertically, at 7 Avon Street. There was a good bit of room in the house; we had three floors, with my study under the eaves on the top floor. The house was heated by a coal furnace that I had to stoke mornings and nights—I seemed to know how to do this. We also enjoyed a backyard, where I built a playhouse for our daughters.

About the time of Cynthia's birth in 1941, I was promoted to tenure as an associate professor. Today the rise in tenure can be a source of considerable academic worry for assistant professors, but I do not recall any such worry in my case. I had spent considerable effort, with Birkhoff, writing the text *A Survey of Modern Algebra*.

Chapter Twenty One ~ Professor at Harvard

Evidently, such action (useful in teaching Harvard undergraduates) did not weigh against me at Harvard (or against Garrett, who was promoted at the same time). It is unfortunate that in today's climate, young mathematicians might not consider writing such texts out of concern that the time taken from their research would negatively affect their chances for promotion.

James B. Conant, a chemist, was then President of Harvard; he was particularly concerned with upholding the quality of Harvard tenure appointments. Consequently, he introduced a system of ad hoc committees to be used whenever Department X wished to make a tenure appointment. The committees, which had outside, non-Harvard members, were to recommend to the Harvard authorities the best people for appointment in the field in question. Conant hoped to counter the usual temptation for senior faculty to support the juniors at hand. At the time, Harvard was open to this trouble because the tutorial system, which required that every undergraduate have a tutor, had brought about many junior appointments in departments with high student enrollment.

The ad hoc committee system brought about an explosion in the Walsh-Sweezey case: both assistant professors in economics came up for tenure at the same time. They were both of liberal political convictions, and perhaps associated with the Communist Party. The ad hoc committees blocked their promotion, and some liberal members of the Harvard faculty suspected that this was because of political reasons. I attended vigorous debates among the faculty about this issue; Stone took a very prominent part of the debates, and was severely critical of Conant. Some older faculty members suggested to me that Marshall might have been able to bring about Conant's defeat if it had not been for distractions from the war.

I was too new to take any part in the debate about Walsh and Sweezey, but my own chief conclusion was that being president of a major university is a difficult job. Conant was cool under fire, and his system of ad hoc committees continues at Harvard.

Many years later, I ran into Conant on vacation in Vermont—to my surprise, he remembered me; university presidents must keep in mind many names and faces.

After the war, I continued to teach modern algebra and sophomore calculus. In the latter course, my wartime experience with airborne fire control enlivened my lectures in elementary differential equations. I also gave a course in statistics—the war had taught me of its importance.

Joseph Walsh normally taught the first-year graduate course on functions of complex variables; it had a central position, coming largely from the recent period when the study of complex variables had been the center of mathematics research. There were also good reasons for this in the beautiful, basic results of this subject. At one point, I persuaded Walsh to swap; he taught modern algebra and I taught complex variables. I much enjoyed this swap, which renewed my acquaintance with the elegant structure of complex variables.

While at Harvard, I guided two more graduate students in their work, and was active in both the AMS and in the Mathematical Association of America. I served as an editor for the *Transactions of the AMS*. I once received a paper purporting to prove the first case of Fermat's Last Theorem, for which I actually managed to line up a distinguished number theorist as a potential referee, though I hesitated to bother him. So I looked at the paper myself and located a total non sequitur ($a = b$ mod 17 does not imply $a = b$), and the paper was rejected. For the MAA, I served as an editor of the Carus monograph series, which then, as now, aimed to publish good expository books. I realized that a good introductory book on logic was missing, so I tried to persuade my friends Steve Kleene and Barkeley Rosser to collaborate on such a book. The joint book never came, but each of them did write an expository book: Kleene wrote *Introduction to Metamathematics*, with a fine exposition of Gödel's theorem; Rosser wrote *Logic for Mathematicians*—a good specialized book is a wonderful thing!

Early in my work at Harvard, I had a confrontation with Bertrand Russell. At the time, he was visiting the United States and one of the social science departments at Harvard. The mathematics colloquium invited him to give an address on foundations, which he did: the audience that came was so large that the colloquium had to move from its regular room to a larger room normally used by physicists.

Chapter Twenty One ~ Professor at Harvard

Russell proceeded to give an enthusiastic lecture, which, roughly speaking, described the state of mathematical logic as it was in 1920. At the end of his talk, the chairman asked for questions. Being a little disappointed that he hadn't covered any recent results, I asked Russell how he related all this to Hilbert's recent work on first order logic and to Kurt Gödel's spectacular results with his incompleteness theorem. There was a long silence. Finally, the chairman said, "Perhaps someone else has a question?" Fortunately, someone did.

I never met Russell again. As an undergraduate at Yale I was mightily impressed by *Principia Mathematica*—Russell had been my hero. By 1938, I realized that new ideas about logic were even more impressive. In retrospect, I felt guilty about my question—I should have known that Russell would not have been able to answer. In doing so, I was too impatient to realize that he had appropriately decided to shift his early interest from logic to other subjects—he hadn't kept up with current results in logic because he had been busy working on other things. However, it is a fact that the splendid background development in *Principia* made Gödel's theorem possible.

Many mathematicians, myself included, were proud of the Harvard colloquium: it presented mathematical results that were both new and important. The audiences were knowledgeable and attentive, with an exception for Norbert Wiener, who sat in the front row and went to sleep yet still asked relevant questions at the end. I recall an occasion when a Russian mathematician (perhaps Gelfond) finished a proof by algebra of an analytical theorem originally due to Wiener, who buttonholed me to ask for the definition of a ring.

My years at Harvard were good ones, but after 10 years, I decided to return to Chicago.

Chapter Twenty Two
University Presidents

By the time I left Harvard, I had watched the work and challenges facing several leading university presidents: Angell at Yale, Hutchins at Chicago, Conant at Harvard, and even Nicholas (Miraculous) Butler at Columbia. Each of them expertly managed systems that were nearly impossible to control. Angell, as an outsider to Yale traditions, somehow managed to convince a conservative teaching faculty to emphasize research; Conant dealt with even deeper traditions at Harvard, but somehow had the charm to facilitate progress. Hutchins tried valiantly to revolutionize general education at Chicago—in the process, he spoke eloquently for academic freedom and managed to gather an enthusiastic student following. And at Columbia, Butler managed to stay beyond his allotted time, perhaps because of the war. However, none of them had it easy.

The difficulties are intrinsic to the position, which involves balancing the budget, encouraging gifts, pleasing the trustees, cajoling the alumni, and keeping both students and faculty in moderate happiness. The president serves many constituencies that, by their nature, are fated to disagree. At Chicago I have seen some presidential failures, and at least one rousing success: Edward Levi handled student unrest at a critical time during the late '60s.

Given my observations of university presidents and my occasional diplomatic ineptness, it is in no way surprising that I never really came close to the work of higher administration; there was only one near miss. In 1949, when I had just moved to Chicago, Charles Seymour resigned as President at Yale. My Uncle John, ever a loyal Yale alumnus, knew committee members engaged in the presidential search, and apparently encouraged them to consider me. I did have an interview, but since I had not yet been department head, much less a dean, I could not have been a serious candidate and did not appear on any short list. This did not worry me; on the contrary, I was much pleased that Uncle John saw that his investment in my education had turned out well—I never did manage to thank him enough. I still have an easy chair from his estate that is said to be made from the wood of the old Yale fence.

It was just as well that I did not become a university president at that time. Soon after my interview came the witch hunt for Communists of university faculties; several universities, Harvard and Yale included, were pressed to reveal and perhaps dismiss members thought to be Communists. Anything of this nature would have been exceedingly difficult for me to handle.

Chapter Twenty Three
Jib and Mainsail

My chief interests in sport have been in skiing and sailing. I learned to ski during my first year at Göttingen, when the Institute for Auslander organized a chance for students to learn skiing in Berchtesgaden; I bought skis and learned a good bit on the hills there, and later skied in the Tirol in Austria. Since ski lifts were not yet commonplace, much of this skiing was old style—we first climbed the mountain and then skied down. Later, I skied in New Hampshire and Colorado, where I managed to sprain my ankle while teaching Cynthia to ski. In later life, I have been restricted to cross-country skiing; now, I can no longer execute a Christie.

Sailing was a much more serious sport for me. Marjorie also owned a cottage in Westport, Connecticut, that was directly on the beach of a bay called Compo Cove. The swimming was good at high tide, which came up nearly to the stone fireplace in front of the cottage. Next-door there was an old tide mill, a millpond, and a millrace. By getting our rowboat up the millrace, I could explore the millpond and its mysterious other outlet on Long Island Sound. One of our relatives gave my family a sail for a rowboat so, when I was a teenager, I carefully fitted out our flat-bottom rowboat with a step for a mast and with sideboards (essentially, these are lee boards that hang over the boat's gunwhale to resist sideways slip while under way). My mother had learned to sail from her father, who learned on catboats in Newport Harbor, Rhode Island. Mother and I sailed happily in and out of Compo Cove, even on the occasion when a strong wind tipped the mast over and pulled out the homemade mast step, which

Harvard Years

Saunders and Gretchen on skis, January 1941

had been only nailed down. Later, I taught my brothers Gerald and David how to sail, and in recent years, Duncan, one of David's sons, has become an expert on catamarans and has sailed in international competitions.

After a while, the older sail rig seemed inadequate, so my brothers and I made—by hand—a new mast, sail, and boom, plus better leeboards. We then sailed around in fine style. Once, in a fresh breeze, David was so concerned for safety that he wanted to jump overboard and swim ashore, but I managed to dissuade him. We sailed on.

After mastering the new sail, I finally decided I wanted something bigger: I bought an 18-foot open cockpit boat, complete with mast, jib, and mainsail, as well as a centerboard (a big improvement on those lee boards). There was even an extra sail. With this, my brothers and I sailed faster and further; I still had the boat in 1944-45, when I was engaged in war research in New York—it provided a happy diversion on occasional weekends.

During my years at Harvard, I did some more substantial sailing with my friend John Cooley, a philosopher I had first met at Yale in a logic course given by F. S. C. Northrop. Cooley came from a sailing family and owned a 40-foot sloop Linnet that he kept in Maine. He was a member of the Philosophy Department at Columbia, but sailing

Chapter Twenty Three ~ Jib and Mainsail

The boat, August 1946

was his real interest, and he often invited me to go cruising with him on the Maine coast. This was fascinating and more serious business; one had to watch the charts and the buoys and listen to the foghorns. At one point, we sailed from Penobscot Bay in Maine over to Halifax Harbor in Nova Scotia; on another trip, philosopher Willard Van Orman Quine came along. Thankfully, we paid more attention to sailing than to philosophical doctrine; I never did much admire Quine's book *New Foundations* (of mathematics)—but I guess I never tried to convert *him* to the categorical foundation of mathematics.

Chapter Twenty Four
Dorothy

For over 40 years, Dorothy suffered from a severe neurological affliction, a variant of encephalitis that became Parkinson's in later life. In spite of resulting constraints, she steadily continued to be outgoing, lively, and courageous—whenever someone inquired about her health, she always answered, "I'm fine." She explained to me that she had been told when she was young that no one was interested in hearing about ailments. She kept welcoming new friends and helping all those around her, and I hope that by describing her life I can illuminate those traits of her character that made her struggles successful.

Dorothy Marsh Jones was born in Governor's Island, North Carolina on July 5, 1907. At that time, her father, Virgil Laurens Jones, from a Tennessee farming family, was superintendent of schools in Faribault, Minnesota. Neither he nor his wife, Isabel Elias, wanted their child to be born a damn Yankee. Consequently, her mother went back to stay with some of her numerous and prominent family in Asheville, where Isabel's father was a lawyer. Both of Dorothy's sisters—Isabel (six years younger) and Alice (ten years younger)—were also born in the South. When each daughter married, their parents made sure that they received a set of silver goblets as a wedding gift so that they could properly entertain dinner guests in the Southern manner.

Dorothy's father had strong academic ambitions, though his education had been irregular; he had no proper high-school diploma. For this reason, Chicago would not accept him as a graduate student, but Harvard would, so he went to Cambridge with his wife and

Harvard Years

Dorothy Jones Mac Lane

infant daughter and finished a Ph.D. in Shakespeare in two years. The next year, he taught at the University of Arkansas and then moved to Sweet Briar College in Virginia. After three years, the family moved back to Fayetteville, where they were soon established in a comfortable frame house right across from the campus and right next to a house later occupied by Hemphill Hosford, Chairman of the Mathematics Department. Virgil rose in the ranks and eventually became Dean of the College; in this capacity, he labored mightily to raise academic standards. Dorothy remembered his particular distaste for those faculty members with the A.B.D. degree—short for All But the Dissertation. Both Virgil and Isabel were suspicious of Bill Fulbright, whose mother had him appointed president of the university (it was later that Bill left for the Senate).

When Dorothy was young, family funds were short, so every penny counted. She recalled, with some pain, that every morning before breakfast she walked a mile to get milk for her baby sisters since there was little refrigeration at the time. Dorothy did well in school, cheerfully skipping grades with her father's encouragement. She rode a bike and played tennis, and had a true friend in Doris Drake from early childhood. Doris especially remembers her kindness; when she lost her mother at the age of 17, Dorothy called on her every afternoon for two weeks. Doris writes, "Everyone appreciated Dorothy's fine mind, strong character, and sense of humor."

Chapter Twenty Four ~ Dorothy

Dorothy spent one year of college at Sweet Briar, an idyllic school for training girls to be Southern ladies. With her favorite camera, she took some striking pictures of the campus and made pocket money selling them. Later, a big, framed copy of one hung on our wall, so that when we went there many years later for me to lecture, I could recognize some places.

For the rest of college, Dorothy returned to the University of Arkansas, where she and Doris organized the first cheerleaders, in red and white outfits, for the Razorbacks football team. Dorothy and Doris were both smitten by religion at about the same time; they made a game of seeing which one of them could call on Reverend and Mrs. Jim Workman most frequently—quite possibly they simply had a crush on him.

Virgil regularly read *The Nation* and had corresponding political views. In 1926, Doris and Dorothy attended an international student conference of the YMCA and YWCA, where Dorothy electrified the audience by proposing, "that we go on record supporting pacifism." It was in the air—that was the same year I wrote my editorial in the high-school newspaper criticizing the military drills we were required to perform. Dorothy admired the YWCA secretary, Fern Babcock, and they remained friends for life.

At about this time, Dorothy learned to drive the family Model T, particularly enjoying gravel roads at full speed. The family went on summer trips every year in that car. Virgil instructed Dorothy on how to recognize respectable hotels—from the sidewalk, you could see into the first floor lobby—and she had good instincts about how to find them. Once, when the family arrived in Los Angeles for the first time, they were at a loss for a hotel, so Dorothy took over the directions: "Turn right here, then left in two blocks." Her father trusted her, and she brought them efficiently to a respectable hotel.

Every trip brought something new and interesting to see, whether it be Bryce and Zion Canyons, Yosemite National Park, or a new town. Dorothy learned that a trip can be fascinating, and all her life she kept on looking, always finding something worth the trip.

In 1926, Dorothy, at not quite 19, graduated from the University of Arkansas. In the sidewalks leading up to the doors of Old Main,

Harvard Years

Anna Dorothy and Dorothy in front of Green Hall, University of Chicago, 1926

where her father had his office, you can see her name inscribed among others of her class. The next fall, her father sent her to the University of Chicago to study for a master's degree in history. She roomed in Green Hall with Anna Dorothy, another enthusiastic Midwesterner. A photo of the time shows Dorothy and Anna Dot in front of Green Hall, with Dorothy standing erect and proud to be there. Dorothy was maid of honor at Anna Dot's wedding to Thomas Wylie, a minister. She and Anna kept in touch through the years; in the spring of 1980, on a lecture tour that brought us close to Kalamazoo, Michigan, Dorothy and I happily stopped off to see Anna, then a widow, in her lovely secluded house on the edge of a small lake.

Dorothy received her M.A. in the summer of 1927, with a thesis "The Record of Eight Democratic and Republican Leaders on the Silver and Tariff Issues, 1878 to 1890." She had learned from Virgil that you had to dig down to the bottom of things, and that this could best be done in Chicago. With degree in hand, Dorothy accepted her first job, a teaching position at Florida State College (now University) in Tallahassee; she was again in the South, young and eager—some of

Chapter Twenty Four ~ Dorothy

Dorothy on one of her many travels, 1934

her students were older than her. However, she only stayed for the academic year. Talking about it later, she left the impression that it hadn't been a successful year; perhaps, after Chicago, it didn't fit her notion of what an effective academic environment really should be.

She spent the next two years (1928-30) in New York City, where she worked at various jobs, finally for a financial house. She lived with several Southern girls in an apartment; one, Edna Stevens, writes me that Dorothy managed the purchase of the food at the market, with the result that the girls were much better fed than their contemporaries. At any rate, that love of markets remained—years later, at our house in Dune Acres, Dorothy loved to visit Jim's Roadside Stand and insisted on picking things out herself, even when her health made that difficult. Perhaps it reminded her of her childhood—Virgil was a farmer's son and Dorothy grew up on vegetables and fruits he grew.

As the Depression made its effects known across the country, Dorothy lost her job at the financial house since she was the last hired. In the fall of 1930, she again came to study at Chicago, this time to take up statistics. This matched her love of specific facts from her master's thesis. She became and remained a member of the American Statistical Association and read its journal; she always delighted in new sources of figures, such as federal statistical tables and the like. Each year, a new World Almanac was the best Christmas present, to be trumped only by two different but competing almanacs.

Harvard Years

Dorothy Jones Mac Lane, 1933

With every new census, our house boasted a new World Atlas, and she gloried in the fact that she had been in every state in the Union—and had a special atlas to count the counties she had visited.

During World War II, while I was off alone in New York, Dorothy worked at one of the wartime agencies in Boston. At that time, she had a dependable housekeeper to care for Gretchen and Cynthia. While working, Dorothy contracted some kind of infection and fell ill for a week. Her regular doctor had joined the war the month before, so could not help or diagnose; the replacement doctor she called told her she had a virus, but he could not do anything to help. In the next few months she began to have episodes of slurred speech and even speech blocks—it was sometimes difficult for her to state what was on her mind. When she came to visit me in New York, we consulted a doctor who told us she needed psychiatric treatment, but that it would be better to wait until we returned to Cambridge. I went to see a psychiatrist recommended to us in Cambridge who suggested that the problem was more neurological than psychiatric, but the neurologist we consulted next did some tests and concluded that Dorothy was normal and not much could be done for her.

Chapter Twenty Four ~ Dorothy

Saunders and Dorothy in Red Square, September 1984

Her problems persisted, and when we moved to Chicago in the fall of 1948, we decided to try the psychiatric approach again. Carl Rogers had recently come to the University of Chicago and agreed to take Dorothy as his patient, but this went nowhere. Then, at an evening party about a year later, one of my colleagues in biology noticed her problem and suggested that neurology was the key. We were put in touch with Dr. Richard Richter, a neurologist at the university hospitals, who diagnosed Dorothy's problems as a variant of encephalitis, probably brought on by the infection she suffered while working during the war. He experimented with medication, which mitigated but did not stop the speech blocks. Dorothy continued to be cheerful to the outside world despite her difficulties in speaking.

Not only was Dorothy's speech impaired, but the onset of Parkinson's and existing arthritis crippled her. In later years she was persuaded to use a wheelchair. Despite this, we managed to travel extensively to places such as Russia and China—she was even wheeled along the Great Wall.

Dorothy had observed the university work of her father as Dean of Arkansas, which helped her to understand the problems that can face faculty members at a university, and she was able to readily assist in many facets of my career. She cheerfully typed up my research papers and book manuscripts. She urged me to go to professional meetings of the mathematical society and to be active there. She was careful to make contacts with other faculty members and their wives, and loved giving dinner parties for my colleagues and our friends. Dorothy instinctively knew how universities work, and so did many things to help me along; she continued to do so despite her own troubles.

We celebrated our 50th wedding anniversary in 1983, with both our daughters and our grandson, William, present.

Chapter Twenty Five
Have Guggenheim, Will Travel

I had profited from my semester research leave from Harvard in 1942, so I applied for a Guggenheim Fellowship for the academic year 1947–48, and it was granted to me. Soon Dorothy and I set off for Europe so that I could learn and do mathematics there. We left Gretchen and Cynthia in Norwalk with my mother Winifred and Marjorie; because of this, we missed some of their early development, but we knew they were in good hands. We sailed to Europe on the Steamship De Grasse.

Paris was our first stop. I had visited Paris briefly during my study in Göttingen, but not enough to really understand the special magic of this city. This time our friend Paul Dubreil, a French algebraist we had met at a conference in Canada, helped us by making a reservation in the Lutetia Hotel, which proclaimed itself to be the Palace of the Left Bank. It was not palatial, but it was on the Left Bank of the Seine where the university and the various other institutes are located, particularly the Institute Henri Poincaré, built for French mathematicians by the Rockefeller Foundation at the same time as the Mathematical Institute in Göttingen.

I managed to give a lecture in my barely adequate French at the Institute Henri Poincaré. After the lecture, one of my auditors came up to introduce himself. He was remarkably cheerful as he reported that the critical reviews of his papers I had written in the *Journal of Symbolic Logic* had been used by French officials to block his promotion. This left me at a loss. I met a number of other French mathematicians, and Marc Krasner, a refugee from Russia. He had acquired all the appropriate French habits, including the importance of shaking hands with each friend at the first meeting of the day; when he arrived late at a lecture, he still took care to shake hands with everybody near his seat, ignoring the lecture. Parisian intellectual life was cosmopolitan, but curious.

There was lively interest in algebraic topology, especially in the work of Jean Leray, whose useful research on partial differential equations had won him a professorship. He was also an officer in the French Army, and had been taken prisoner early in WWII; he was interned in the Prison Camp for Officers in Vienna. Since he was also a full professor, the Germans made him president of the university in the prison. He continued his research from prison, but apparently could not get access to the current literature of his field of specialty in partial differential equations, so he shifted to topology, where he invented two powerful new devices, sheaves and spectral sequences. Both were important for the future in topology and in my own future research.

I first listened to Leray's lectures at the College de France, where he had just been appointed to a professorship. Professors there were required to give each year a series of lectures on their specialty. In the year I was there, Paris was cold and some of the attendees perhaps came just to warm up, but there were better reasons; two students, J. P. Serre and Armand Borel, asked their mentor, Henri Cartan, whether Leray's lectures were worth attending, and he said they were. After a couple of weeks, Serre gave up, mystified, but Borel continued until the magical point where he began to understand. He then found Serre, to tell him, "Look, Jean-Pierre, you can actually use those strange spectral sequences to prove real theorems." Two weeks later, the *Comptes Rendues* carried an article by Borel and Serre that used spectral sequences to prove that a Euclidean space cannot be fibered by compact fibers (that is, by spheres). After that, Borel and Serre went on to write their respective and impressive theses, using spectral sequences to study aspects of Lie groups (Borel) and the hard problem of calculating homotopy groups of spheres (Serre). Serre's thesis made topologists sit up and notice spectral sequences.

Chapter Twenty Five ❧ Have Guggenheim, Will Travel

My own spectral sequences came more slowly. I listened to Leray, but didn't understand, and at the time, did not even see that the thesis of my Harvard student Lyndon was actually a use of spectral sequences. Lyndon was then working in London and visited us in Paris, but we didn't make the connection at the time. Much later, in Chicago, I realized that I had to understand this strange engine; I first gave a quarter course on spectral sequences, had a student write up spectral notes, and then directed his doctoral thesis on their properties. I finally organized a systematic presentation in my book *Homology*. In this case, it took a long time for the idea to shake down, but it is now essential in algebraic topology.

In coming to Paris, we brought some food for a friend, Mademoiselle Lehnaurdie, because the country still suffered from the shortages and the destruction of the war; indeed, many people lived in damaged apartments. She invited us to dinner at her home, where she lived with her sister; it was a rickety studio apartment on the fifth floor of a new apartment building that in New York would have been called a cold-water flat. The apartment had a big studio, a miniscule kitchen, and a living/dining/bedroom. The apartment was once well-heated, but at that time, the only source of heat was the fireplace—perhaps it was because of a strike or a gas shortage after the war; Paris was in a confused state in those days. In any case, the gas pressure was very low when supper was to be cooked, so a special gas burner was used that had the special property of increasing the pressure at the danger of suffocation. In spite of the disadvantages, we enjoyed a very delicious meal.

After our brief stay in Paris, Dorothy and I went on for a longer stay in Zürich for the chief reason of visiting Heinz Hopf, a professor at the Swiss Federal Technical Institute—he was the topologist whose ideas started me and Eilenberg working on the cohomology of groups. While we were there, I listened to Hopf lecture on differential geometry, and I also learned from his students, in particular from Beno Eckmann. We enjoyed our life in Zürich, with the beauties of the lake, but we did not really succeed in understanding the Swiss. We lived well in a pension, Zürichburg, on top of a hill overlooking

Dorothy with Heinz Hopf, Zürich, 1954

the university. We usually got uphill to our pension by streetcars, which were tiny but powerful in Zürich; the fare was determined by the number of sections traveled, which was hard for us to compute because we did not know where the sections started and ended.

The center of Zürich is impressive, built around the river Limmat, which flows north from the lake of Zürich. While there, we visited the charming medieval town of Schaffhausen on the Rhine; Lucerne, though we did not find the Lion of Lucerne; and Basel, where I gave a lecture and met several Basel mathematicians, including A. Speiser (a group theorist) and A. Ostrowsky, whose papers I had studied carefully.

I also decided to visit Germany once again—it was in poor condition then, divided up into French, British, U.S., and Russian zones. I had to get official permission to visit. First, I traveled to Göttingen to visit the university where I had so seriously studied. The faculty members from my time were, as I knew, dispersed; however, Professor Herglotz was still in office, though in poor health. Somewhat to my surprise, he remembered me, and I guessed he had pleasure recalling his earlier mathematics triumphs. The Mathematical Institute was open and functioning, and the well-stocked library still had available the thesis that required a trunk to hold it; it was an earlier thesis that had exemplified a contentious construction by ruler and compass. I also met various new, younger faculty members, such as Wilhelm Magnus, Arnold Schmidt, and Franz Rellich. Helmut

Chapter Twenty Five ~ Have Guggenheim, Will Travel

Hasse was still at Göttingen, but was out of town at the time. It was interesting for me to return to the university, but my visit did not (and could not) recapture the older magic of my studies there.

Göttingen was practically undamaged during the war; a local Nazi leader, on a Saturday, proclaimed in the papers that Göttingen would be defended by the Volksturm—unarmed civilians—to the last man. He left in a hurry Sunday morning, just before the American troops arrived. Except for R. Cauer, in whose apartment I had lived when I first came to study there, all the local mathematicians came through unscathed. On the day after the occupation, Cauer insisted on returning by bicycle to his wartime job in Berlin, where he was shot as a hostage by the Russians.

While there, I gave three lectures; one each on algebra, topology, and current literature, since the mathematics there was out of touch. I also distributed various supplies I had brought from Switzerland, such as butter; at that time, Germans had trouble getting items, often standing in lines to get a permit, and in another line to get it stamped.

On the way back from Göttingen I stopped in Frankfurt, where I met the topologist Wolfgang Franz and his wife, who was very active in rebuilding and reorganizing bombed-out apartment houses, her own included—the destruction was still evident.

After Frankfurt, I stopped in Heidelberg, which was in the American zone, so I was lodged in one of the elegant hotels then managed by the American Army. Fortunately, Heidelberg had been little damaged by the war. The elegant old bridge was still there, running across the river Neckar. And that vigorous publisher of mathematics, Springer-Verlag, was still located in a big building on the north bank of the river with a good view of the castle. Every spring, there was a fireworks display at the castle, and the old Springer office had an excellent view. Springer now has a newer but less romantic office building far from the view of the old city with its famous castle.

Ferdinand Springer attended one of my lectures and later invited me to his home where he recounted an interesting war story. Just before the end of the war, the Russians had captured him, and the major who interviewed him asked him who he was. "I am a scientific publisher. I publish about a hundred journals." He was told to write down the titles. When he reached title number 90, the major said, "That is enough; I have published papers in this journal and in this one." It appeared he was a geneticist; in fact, he had apparently been an editor of one of Springer's journals. Springer had refused, despite a request from the Nazi government, to remove the editor's name. After two days, the Russian was willing to let Springer go, but advised him to stay with them, lest he be captured again by a less-scientific Russian unit.

At the time of my visit, 1948, the dominant textbook on topology was the well-known Seifert-Threllfall *Einführung in die Topologie*. It was noted for the clarity of its explanation of homology and for its splendid, more intuitive discussion of covering spaces and their relation to subgroups of the fundamental group. The authors of this marvelous text were both professors in Heidelberg, and they received my visit with enthusiasm and introduced me to their new Ph.D. student, Horst Schubert, who later became an expert on categories. Topology was then (and later) very much alive in Heidelberg.

At that time, Seifert and Threllfall worked out of a small office in an old university building and lived in the university observatory, the Sternwarte, which was on the top of the small mountain south of the town, served by a mountain railroad that initiated in the town. Dorothy and I enjoyed our visit there, and I treasured the intellectual activity in Heidelberg.

In 1948, Heidelberg offered a strange contrast between two completely separate societies: the shabby and unhappy Germans, and the well-dressed, seemingly unaware Americans. The Americans had taken over all the best hotels, as well as the good restaurants; there, the people in possession of script—military payment certificates—lived. They had American meals, coffee for breakfast, the *Paris Herald*, movie magazines, Camel cigarettes, and obsequious German waiters who spoke perfect English. There were bars, nightclubs, ice cream parlors complete with Coca-Cola, and a PX with Ivory soap, Planter's peanuts, and prophylactics. It was just as if Emporia, Kansas, were superimposed on Heidelberg, but with little mutual contact as if they were separated by their normal geographical distance.

Chapter Twenty Five ⁓ Have Guggenheim, Will Travel

There was similar contrast in Göttingen, where the British officers of the Occupation enjoyed hunting wild boar. But the town of Göttingen had not lost its charm: The City Hall, where Dorothy and I had been married, was still handsome, and still had the statue of the Gansemädl, the Goose Girl, which was known for the legend that students must climb up to kiss her after passing the final exam (I had skipped this). One could still walk agreeably around the town on the top of the old, earthen town wall. In front of one stretch near the Math Institute was the Gauss-Weber Denkmal, which showed Gauss and Weber in the midst of their famous collaboration on the theory of electromagnetism: Weber was seated, with Gauss standing at his shoulder. Student legend had it that Gauss was saying to Weber, "Get up, old chap, and let me sit down for a change."

Heidelberg was, and still is, an even more romantic setting. One can walk high up along the right bank of the river along the Philosophen Way, which allows one to think as one wishes of philosophical issues or the beauty of the town. The philosophical issues are surely present: Heidegger was once professor there, and I fondly imagine that it was there that he concocted some of his jaw-breaking ideas. I cite an example from his book on Nothingness, *Nothing Nothings*. It is more profound in the original German, *Das Nichts Nichtet*. Or, if while on the splendid path, one does not wish to settle philosophical issues, one can look out over the town dominated by the romantic old castle. In some past century, the French invaded and bombed the castle. Some of the old walls have been left in their bombed-out state, which makes a striking romantic view and a small indication of the trouble with having a castle that can be bombed.

The unbombed section of the castle has a courtyard where theater is presented; on a subsequent visit to Heidelberg, Dorothy and I had dinner in a castle room overlooking the courtyard, and from there, viewed a performance of "The Student Prince." It was easy to imagine oneself as a student and a prince, busy attending the ancient university—it outdoes the Yale fence.

The rest of the winter in Zürich went well. We attended the opera, and I went skiing at Saint Moritz in the Alps with Richard Buchi, a young logician, who later moved to Purdue; many years after that, I served as an editor for his collected papers. I had a good opportunity to talk about logic with my erstwhile Göttingen mentor, Paul Bernays, whose wide knowledge of logic was most impressive. I was invited to attend the 400th anniversary of the Charles University in Prague, but the invitation was canceled when the Communists took over in Czechoslovakia.

After Zürich, Dorothy and I traveled to the Netherlands and Belgium (to visit topologist Guy Hirsch) and then on to Great Britain to visit both Cambridge and Oxford. At Cambridge, I finally met Philip Hall, the algebraist and group theorist who had so influenced Garrett Birkhoff. I even tried, with little success, to punt on the River Cam. I also listened to some of the Cambridge views of algebraic geometry; they struck me as being decidedly old-fashioned.

In Oxford, British topologist J. Henry Whitehead met and entertained us. After his undergraduate work, Henry had made money on the British stock exchange and then took up topology. At that time, there was little up-to-date research in topology in Great Britain, so Henry went to Princeton and wrote his Ph.D., guided by Oswald Veblen. With Veblen, he wrote the first serious attempt to define differential manifolds in general terms (the subsequent definitive definition was later provided by Hassler Whitney). For topological purposes, Whitehead later introduced the useful notion of a CW-complex, a space put together in a proper way from basic Euclidean cells. J.H.C. Whitehead wrote vigorously, but not always clearly; hence, he earned the nickname Jesus He's Confusing.

He had spent much time analyzing the homotopy type of spaces: two spaces X and Y have the same homotopy type if there are continuous mappings $f: X \to Y$ and $g: Y \to X$ such that both composites are homotopic to the identity map in question. Would the usual homology groups be sufficient to determine the homotopy type? No. Were the homotopy groups enough? Again, the answer is no. Whitehead probably knew this already, but he thought that the recent results by Eilenberg and myself would help.

Chapter Twenty Five — Have Guggenheim, Will Travel

What Whitehead and I found was the definition of a new invariant $k^3 \in H^3(\pi_1, \pi_2)$, relating the fundamental group π_1 and the second homotopy group π_2 of a space. This invariant is a cohomology class that uses the way that π_1 acts on π_2. We proved that, when all higher homotopy groups vanish, the homotopy type of the space is determined by these three invariants: π_1, π_2, k^3. This result made essential use of Eilenberg's singular homology theory and of the Eilenberg-Mac Lane cohomology of groups. Whitehead and I published our result in the *Proceedings of the National Academy of Sciences*.

This result was just a first step in a possible more extensive description of homotopy type. If the next homotopy group π_3 is nonzero, a cohomology class $k^4 \in H^4(\pi_1, \pi_2, k^3, \pi_3)$ can be defined, which is an invariant. These homotopy groups π_1, π_2, ... and k invariants k^3, k^4, ... in effect provide a complete description of homotopy type. They form what is now called a Postnikov system because the Russian mathematician Postnikov first formulated it in print. But the description of such a system can, I now suspect, be put into more effective algebraic forms, as has been tentatively suggested in unpublished research by Joyal and Tierney.

At the time, Henry Whitehead and I had tentative plans to do further joint research, but they never came to fruition, largely due to my negligence. Whitehead made several visits to the United States; at one point, he stopped off to visit his uncle, the famous philosopher Alfred North Whitehead, who was then a professor at Harvard. In the course of a weekend visit, he chanced to fall into some disagreement with his uncle. As he prepared for his train journey, his aunt tried to cheer him up: "Henry dear, do you have something light to read on the train?" she asked. "Yes, Auntie," came his reply. "I have a copy of Uncle Alfred's *Process and Reality*."

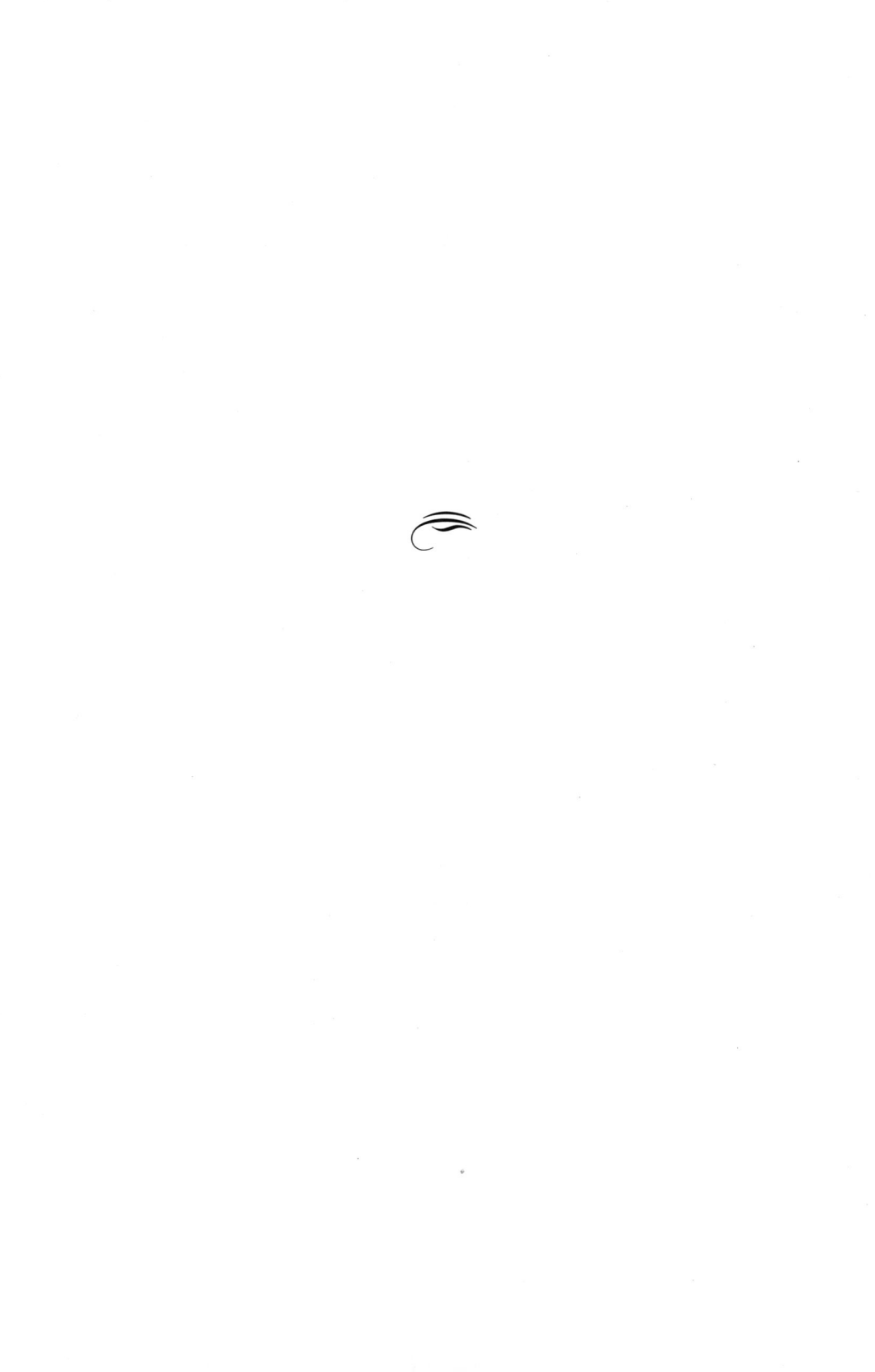

Part Seven

Chicago in the Fifties

Chapter Twenty Six
Return to Chicago

In the fall of 1948, Dorothy and I returned happily from our year in Europe, heading straight to Norwalk to find Gretchen and Cynthia in good spirits. But we did not go back to Harvard—before leaving for my Guggenheim Fellowship, I had accepted an appointment at Chicago. There were two reasons for my decision: the lure of new mathematics at Chicago and Dorothy's discontent with the traditional atmosphere in Cambridge.

Mathematics was undergoing changes at Chicago: the Manhattan Project for the atom bomb had been established there, with headquarters in Eckhart Hall, the mathematics building. With the coming of the war, the university decided to continue this type of organized research in physics and chemistry by setting up two research institutes at Chicago, now called the Enrico Fermi and the James Frank Institutes. But it was thought that the institutes would require more mathematical support; it was clear to the Chicago administration that the long-existing mathematics department, chaired by Gilbert Ames Bliss and then Ernest P. Lane, had misplaced its emphasis, as I myself had realized back in 1931. So President Hutchins asked John von Neumann, who was well known for his work with the Manhattan Project, to come to Chicago to reinvigorate mathematics. He declined the offer, but suggested that Chicago approach Marshall Stone of Harvard instead. Stone and von Neumann had worked in parallel earlier on the very active development of Hilbert spaces and the spectral theorem, both for the development itself and for its amazing applications to quantum mechanics.

Chicago in the Fifties

Sammy Eilenberg, 1950

During the war, Stone was engaged in some cloak-and-dagger operations in India. He had been active in Harvard mathematics, but evidently felt that the Harvard administration was both too conservative and too centralized to be aware of the growing needs of mathematics. Put differently, Stone foresaw that with the end of the war and with the many talented European mathematicians now sheltered in the United States, there was a real window of opportunity to bring about a major advance in American mathematics. In the end, he was persuaded that Chicago could be an ideal location for such a mathematical renewal, so he resigned from Harvard and began to seek mathematicians to bring to the new venture at Chicago.

He had administrative support at Chicago; however, that support was not always sufficient or, perhaps, not always wise. For example, he knew that André Weil was a dynamic mathematician who had not been well-placed—when Weil had first come to the United States during the war, he had been put to teaching calculus at Lehigh, and subsequently left for a better position in South America. Stone asked the Chicago authorities to make Weil a suitable offer. As soon as Stone learned of the financial terms of the offer, he at once wrote Weil that the terms were not adequate, and that he should decline. Weil did decline, and a more appropriate offer came later, which he accepted. There were probably other problems in assembling the new Chicago department. The only one I definitely know is that Stone asked Hassler Whitney to come to Chicago from Harvard. Whitney declined, but he did not stay long at Harvard, leaving instead for a

Chapter Twenty Six — Return to Chicago

professorship at the Institute for Advanced Study. While there, Whitney changed his interests from algebraic topology to developing better mathematics teaching for young children.

Early in 1947, just before I left for Europe, Stone asked me to come to Chicago, telling me that President Hutchins also wanted me to come. At Marshall's suggestion, I came to Washington to discuss the situation with him in the home of his father, Harlan F. Stone, who was then a member of the Supreme Court. During the interview, Marshall was enthusiastic about the prospects of a drastic renovation of mathematics to be built at Chicago—his insights were incisive, and I shared his view that the Harvard administration was too conservative.

When I returned to discuss the prospects with Dorothy, Chicago was attractive to her because of her earlier studies there, as well as our one-year stay there together. Also, Dorothy had never really fit in well with the traditions or autocracy in New England. Moreover, the Chicago offer was exciting as something new, more vital, and more innovative than seemed possible in Cambridge. So, after negotiating with a dean, I accepted the Chicago offer, with an arrangement for continuation and partial funding of my already planned Guggenheim year abroad. I had no real complaint with Harvard, but we decided that when we returned from Europe, we would go to Chicago.

I have since visited Harvard and Cambridge from time to time, once as a member of an advisory committee to the Department of Mathematics, but the institution seemed very distant to me, almost as if I had never lived there.

Chapter Twenty Seven
The Stone Age at Chicago

By good instincts and foresight, Marshall Stone succeeded in creating an innovative and inspiring new department of mathematics at Chicago, devised by a happy combination of an international senior faculty, an ambitious junior faculty, and an unusually lively group of graduate students. For a period, it was arguably the best department in the world.

The senior faculty members were Adrian Albert, S. S. Chern, Marshall Stone, André Weil, Antoni Zygmund, and myself. Albert was a holdover from the old Chicago department, but Stone brought in Chern, me, Weil, and Zygmund, a remarkable quartet of senior appointments in such a short time that would have given a big shot in the arm to any department, and it certainly did so to Chicago.

Albert had started as a student of Leonard Dickson in algebra; he was aware of the nature and troubles of the old Bliss department and was happy to join Stone in the new direction. His research interests continued some of Dickson's in group theory, linear algebras, and nonassociative algebras.

Chern, from China, had studied in Europe with Blaschke and Elie Cartan. He knew and understood the strength of Cartan's use of differential forms in geometry.

Stone, as already noted, led in the use of Hilbert space methods in analysis and physics. The new department was definitely his design—he was, in effect, a dictator, not just a chairman.

Weil had grown up in the demanding and high-reaching world of French mathematics. He had found decisive results in algebraic geometry, and he and his young colleagues in France had realized the importance of the abstract methods newly used in Göttingen. They started the influential Bourbaki group in the ambitious project of an exposition of all of basic mathematics.

Zygmund and his wife had escaped Poland at the start of the war. Between the wars, Polish mathematicians established a lively school of mathematics, deliberately restricting their research interests to chosen topics in topology, logic, and analysis. Zygmund was a devoted specialist in analysis, particularly in harmonic analysis. He was equally devoted to his many students. Other refugee mathematicians from Poland displayed a similar devotion to their own specialties.

Stone's junior faculty, who all began as assistant professors, consisted of Paul Halmos, Irving Kaplansky, Irving Segal, and Edwin Spanier. Halmos, a Hungarian, was a protégé of von Neumann, and an enthusiastic expositor on vector space theory. Kaplansky, my first Ph.D. student, had met Albert during the war, and actually had come to Chicago on Albert's initiative a year before Stone arrived. Segal, another von Neumann protégé, studied at Princeton and taught briefly at Harvard. Spanier, a recent Princeton student in topology, had learned the flourishing new ideas of algebraic topology from Steenrod.

At the start of the Stone Age, Otto Schilling was still in Chicago; his mathematical interests in algebra had been interrupted by legal problems involved in a troubled financial inheritance from Germany—my collaboration with him sadly came to an end. Eventually, he moved to Purdue University. A few years later, I. N. Herstein, an algebraist, joined the department, and several of the younger mathematicians from the Bliss-Lane department left. Professor Lawrence Graves retired, as did H. S. Everett, who had effectively managed correspondence courses in mathematics.

With the start of the Stone department, many new, able graduate students arrived at Chicago. A number of them were supported by the GI Bill, which provided adequate stipends, and many students came because they had used the Chicago correspondence courses prepared by Professor Everett during their war service. Other students in Urbana or in New York City heard of exciting things going on in

Chapter Twenty Seven ~ The Stone Age at Chicago

Chicago, and so came to study there, and there were also several fellows who came from Europe to visit Chicago. One British visitor returned to England to build up a Chicago-style department in Warwick.

As chairman, Stone was the leader and manager; he was forceful and judicious—perhaps growing up as a son of a chief justice of the Supreme Court helped. The story goes that at one point, he was busy in his offices while a student, Bert Kostant, waited patiently at the door. Professor Weil came to see Stone on some matter, opened the office door, and walked in, whereupon Stone pointed out that the student had been there first.

Stone wholly reorganized the graduate program in mathematics; previously, the final Ph.D. examination formally covered all 27 graduate courses the student had taken. I recall one example where a student had taken all 27 courses, and finally, in absentia, finished his thesis and come back for his final exam. One of his courses had been algebraic topology, and the examiner asked the student for the definition of a covering space; at first, the student was stumped, but on further questioning he came up with an example, the line and the circle. The examiner probed further to ask which covered which, and the student thought that the circle covered the line. Disaster! The solution is to wind the line again and again around the circle, which makes the line cover the circle.

In the reorganized Ph.D. program, the final Ph.D. oral exam no longer probed all the courses the student had taken. Instead, it stuck to the thesis and related things, and was pretty much a formality, since the student had already shown his or her breadth of knowledge in the new Ph.D. qualifying orals. Under this new setup, the situation described above might not have happened and the student would have passed.

In the Stone department, there was a whole new sequence of required graduate courses. After the master's courses, there were no further required courses for the Ph.D.; the student chose advanced post-master's courses, took qualifying exams, and completed his or her research. In many cases, faculty prepared mimeographed notes for the new master's courses: Zygmund prepared notes on analysis;

others prepared notes on set theory and metric spaces, as well as several alternative sets of notes on point-set topology (one of which was mine). The latter notes recognized that topology had a central position, comparable to that once held by variable theory. The Stone-Weierstrass theorem made a vital appearance there. Weil complained that the classical definition of determinants should be replaced by a definition making some use of Grassman algebras, which I presented in some lecture notes. The master's program represented a new, systematic view of mathematics.

This new model of emphasis cropped up all over, and there were new advanced courses. Weil's book on the foundation of algebraic geometry appeared in the Colloquium Series of the AMS. He lectured on the subject; I read and listened, and in my reading, I noted that this version of algebraic geometry could make considerable use of category, an observation that I regrettably did not follow up (later, Grothendieck developed the same observation). Weil also conducted a seminar on current literature, inspired by an older Parisian seminar taught by Hadamard. The principle of the seminar was that each student should report on a current paper *not* in his field of specialty; the student reports would be open to criticism. It is interesting that Weil saved his most devastating comments for those students he knew to be the ablest.

Zygmund taught analysis and encouraged many notable students, such as Alberto Calderon and Elias Stein. He impressed upon all of them the importance of harmonic analysis. Chern taught differential geometry using differential forms, of course. Isadore Singer and other students learned from this approach, which was not yet readily available in texts. It is reported that at one time a student told Weil that he (the student) did not understand these differential forms, to which Weil responded by going to the blackboard and writing down the Greek letter omega (ω); I suppose he intended that this standard symbol for a differential form might recall the idea. Spanier and I alternately taught algebraic topology; in particular, I struggled with the still-mysterious properties of those spectral sequences found by Leray and now used by all topologists.

Mathematical discussions continued in the corridor and at the daily tea. For example, I recall trying to persuade some students at tea

Chapter Twenty Seven ~ The Stone Age at Chicago

of the logical existence of a maximal atlas for any differential manifold—logic was not, as before and afterwards, isolated from mathematics. In this hothouse atmosphere, ideas and proofs were prominent and many graduate students flourished, including Isadore Singer, Bert Kostant, Richard Kadison, John Thompson, and many others.

As an illustration, I will describe Thompson's work: He had been an outstanding undergraduate in mathematics at Yale, and the reputation of the new department probably attracted him to Chicago for graduate study. Shortly after his arrival, I wanted a chance to do something different, and decided that group theory was due for a revival; hence, I taught a two-quarter course on group theory. By the end of the second quarter, in which I had tended to emphasize infinite groups, I had essentially come to the end of my knowledge on group theory. Thompson, who was in the course, came to me to say that he wished to write a thesis on group theory. I encouraged him, but did not trouble to say that my own knowledge of the subject was somewhat limited. But not to worry—I arranged for eminent theorists, such as Richard Brauer, Reinhold Baer, and Marshall Hall, to visit Chicago. Each Saturday morning, I listened to Thompson tell me what he had been up to with groups; the subject fitted his interest, and he chose his own problems.

At the time, André Weil was lecturing on group representations with particular attention to the so-called Chevalley groups. One of Weil's students heard what Thompson was up to and took him aside to tell him that he should not be studying finite groups alone, but should look at representations as well. However, Thompson persisted, and turned out a Ph.D. thesis that settled an outstanding problem of finite groups, a construction of a normal p-complement for certain finite groups.

Adrian Albert was also interested in finite group theory, a subject active at Chicago from the very beginning in 1892, and found special funding to organize a special year on group theory at Chicago. Thompson and Walter Feit, a recent Ph.D. of Richard Brauer at Michigan, were two of the participants; during the year, they solved a famous problem by proving the Odd Order Theorem: A finite simple group that is not cyclic cannot be of odd order, which was a famous conjecture of Burnside. This result was the starting point of a major effort to classify all finite simple groups.

Group theory, as here exemplified, was just one of the topics that flourished during the Stone Age at Chicago. The people involved—the students and professors—were excited by the pursuit of new ideas in mathematics, which was, in all, a happy combination of talent, circumstance, and leadership.

Chapter Twenty Eight
The Stone Age Comes to an End

Marshall Stone served as chairman at Chicago until 1952. By this time, he was very weary of what was a continual struggle with people in the administration; perhaps his ideals were so high that it was hard for him to take a realistic course. But his interests had turned toward international issues, particularly the International Mathematical Union, which had gone out of existence at the end of World War I because of its troubles with the questions about the participation of Germany. Stone took the lead in organizing a new Union, which was essentially launched at an initial meeting in Rome in 1952. I was actually one of the American delegates to this meeting; I enjoyed a luxurious trip over to Italy on a fast new American ocean liner (the SS United States). The new union was organized with spirit and enthusiasm. I brought Dorothy some elegant new Italian china; though Stone was the president of the new union, he failed to bring home the china his wife had hoped for.

When Stone ended his term as chairman, the usual process of selecting a replacement went on in administrative secrecy (the department would later change this process). I was asked to accept the position, but was a bit hesitant about taking on administrative work because of the inevitable loss of research time. However, I felt that I understood what Stone designed, so I accepted. I moved into the chairman's office, and Stone moved into a more secluded spot. With his international interests, he traveled widely and so had little direct influence on department affairs, and I endeavored to continue in the direction he had established.

Saunders representing the U.S.A. at the International Mathematical Union in Rome, March 1952

The faculty was still excellent and mathematical ideas still prospered. But the chairman's job was also a troubled one, with its major policy issues and assorted bureaucratic regulations. In 1954, it happened that Felix Browder came to give a colloquium lecture on partial differential equations. It was a good lecture; Weil pointed out to me that this subject was one that was not very well represented at Chicago. At that time, possible departmental appointments were handled administratively, often without discussion at department meetings (of which there were then few). At any rate, I proposed to the administration that we offer an appointment to Felix Browder as assistant professor.

Hutchins had left Chicago, and the new president was Lawrence A. Kimpton, and our proposal to appoint Felix Browder was not approved. It was perfectly clear why, though the reason was not explicitly stated: Browder, though not himself politically active, was the son of Earl Browder, who was recently the head of the Communist Party of the U.S.A. This came at a time when there was considerable anti-communist agitation. I felt that such considerations were out of place at a university, and was tempted to resign from the chairmanship,

Chapter Twenty Eight ⁃ The Stone Age Comes to an End

but Stone persuaded me not to do so. We did not propose another appointment. Browder later found a position at Yale, but eventually did join our department (1963), and even later served as chairman (1972–77 and 1980–85).

I experienced various administrative troubles as chairman; for example, it was hard to keep an efficient department secretary. The personnel office did not seem to understand our needs—at one point, they sent me a form on which to rate the performance of the department secretary, which I converted into a rating of the personnel office and returned to them. I reported chiefly to the dean of the Division of Physical Sciences, who was supportive. His secretary, however, was not—she took over, for her use, a storage closet right there on the mathematics floor of Eckhart Hall.

I also had some administrative work for the International Mathematical Union, where I was a member of the Executive Committee. As part of this work, I helped the Union organize a small international conference, in Germany, on mathematical logic, which went well. The Union was succeeding, though not without minor conflicts; I recall some sharp disagreements with W.V.D. Hodge, who represented the United Kingdom.

The Union did succeed in organizing the International Congress of Mathematicians, which occurred every four years. I was pleased that one Congress took place in Edinburgh, coupled with a cruise to the Inner Hebrides. On the cruise, Serge Lang got several people to swap their identification badges (Lang himself could be recognized without a badge); this was typical of Lang—he liked to make jokes. Some years after the Edinburgh conference, there was another in Amsterdam. I enjoyed the lectures, but walking through the town with my family, we were almost hit by a car. Speaking in German, I protested, and landed in jail. It took Stone to get me out.

In 1956, the redoubtable John von Neumann, a long-time professor at the Institute for Advanced Study, died. Shortly thereafter, the Institute offered Weil a position as professor. At that time, he was on vacation in Paris, staying at his family's apartment. It happened that I was soon to travel to Paris for a meeting of the Council of the International Mathematics Union, and would be able to meet with him. I realized that Weil was, in a real sense, the leading member of our department, and that the offer would tempt him. So I made an appointment to see President Kimpton and asked his support for a suitable counteroffer to Weil. I described his decisive contribution to our department, and Kimpton gave me nothing to offer Weil—no raise in salary, no distinguished title, nothing. Perhaps I should have made my appeal for help through the regular channel with the dean of the Division of Physical Sciences, but I doubt if that would have made any difference. It is unlikely that a good counteroffer would have kept Weil at Chicago; just recently, I happened to see a statement he had written before this event to the effect that a position as professor at the IAS was the top position for a mathematician in the United States. Predictably, Weil accepted the offer from the IAS, and I consulted department members as to possible action.

We received permission to approach a German mathematician, Fritz Hirzebruch, with an offer to join the department. Hirzebruch had introduced important new ideas in topology and had spent considerable time at Princeton, but by the time of our offer, he was already a professor at Bonn in Germany. Despite my appeal to him, he was pleased with his position there and declined to come. I did not succeed in filling Weil's vacancy.

I served two three-year terms (1952-1958) as chairman; during the second term, I took a leave of absence for a year, during which Adrian Albert served as acting chairman. In the total six-year period, I did not succeed in making any new tenure appointment to the department. I of course realized that such a situation is destructive for a department with major aspirations. Otherwise, I did manage to hold the department together and keep it on course, except for the loss of Weil. At one point, Paul Halmos, then an associate professor, received an outside offer; although I did not then recommend an increase in salary for him, he did stay in Chicago.

I did succeed in making some junior, nontenured appointments. There were several able young mathematicians, such as Serge Lang, who came as instructors. Lang was notable among them, in part because he declined to recognize the Chicago weather—he bought no overcoat. At one point, the department decided to do better by

Chapter Twenty Eight ~ The Stone Age Comes to an End

our calculus students by having the calculus classes meet five days a week instead of three. Lang firmly declined; in three days a week he could teach them everything that they possibly could learn. Lang later acquired a resounding reputation as the best mathematics faculty member at discussing administrative nonsense. Could Chicago claim any credit for training him to recognize and speak up when confronted with buncombe? Lang has become an expert on such matters, such as denouncing government agency rules that faculty account for percentages of their time.

As chairman I also managed to make several promising appointments of assistant professors: Walter Baily, an expert on class field theory from Princeton; Eldon Dyer, who later moved to CUNY; Shlomo Sternberg, who later moved to Harvard; and Sigurdur Helgason from Iceland, who later moved to MIT. In short, my work as chairman was essentially a prolongation of the Stone Age that continued, but did not expand, that tradition.

Adrian Albert became the next chairman in 1958. Another serious loss arrived quickly after he took office: the University of California at Berkeley offered positions to Chern and Spanier. The offer was tempting to Chern because of the considerable concentration of former Chinese citizens in California. Despite Albert's efforts, both Chern and Spanier left for Berkeley. In sequel to the loss of Weil, this was, in effect, the end of the great Stone Age department. Stone himself took early retirement and moved to a position at the University of Massachusetts in Amherst, where he continued his long-time interests in the structure of mathematics. At one point, he returned to Chicago as a visitor to lecture on such topics.

Albert made various attempts to rebuild the department; at one point, he made an attractive offer to Samuel Eilenberg, who used it as a lever to build up his department at Columbia. Thus, the Stone Age of Mathematics at Chicago came sadly to an end; but the department continued in its pursuits.

Chapter Twenty Nine — Hutchins and the University

In 1929, Robert Maynard Hutchins became President of the University of Chicago at the age of 30. His administration there exerted a decisive influence on all of American higher education. He insisted that universities should attend to the basic aims of education and took courageous stands defending academic freedom. After WWII, he actively revived and strengthened the sciences at Chicago. Finally, and most characteristically, he emphasized the central importance of the Great Books, and established general education in a remarkable new college at the university.

Hutchins studied for two years at Oberlin College and then served as an ambulance driver in Italy during WWI. Afterwards, he finished his undergraduate degree at Yale in 1921, with a beginning study of law. He spent two years as a master for English and history at a private school in Lake Placid, New York, and then returned to Yale and completed studies for the L.L.B. at the Law School, while simultaneously serving under President Angell as Secretary of the University. At the Law School, he was a lecturer from 1925–1927, a professor in 1927, and dean from 1928–29. As dean, he advocated broadening the rather narrow scope of legal education; in particular, he wished to connect legal training with studies in the social sciences. After examining his career path, it is easy to see that Hutchins's university background was more on the educational and administrative side than in traditional scholarship—this reflected in his style as a university president.

The Hutchins College developed out of an earlier interest in general education at Chicago. New plans in this direction were already at hand when Hutchins became president; in particular, the graduate faculty had organized new survey courses, in areas such as the physical and biological sciences, for general studies. Hutchins drew on the practice of general education at Columbia through his friend Mortimer J. Adler, who had studied there and then worked with Hutchins at Yale. Much as he had invited me to Chicago as a graduate student, Hutchins appointed Adler to an associate professorship in philosophy at Chicago without consulting the senior members of the department. He was indeed an autocrat.

Initially, Adler and Hutchins taught courses on the approximately 100 Great Books, emphasizing the Great Ideas they present. Some of these books, such as Newton's *Principia*, must have seemed obscure to the students, but those books were (and are) a fortunate alternative to the common practice of teaching from standardized textbooks, where the Great Ideas may not seem so great because they are often diluted, condensed, and trivialized.

However, Hutchins still wanted a more inclusive general education for all undergraduates, an education that was not under the control of the powerful subject-matter specialists of the Chicago graduate departments. He quickly appointed a couple of his supporters to the faculty, and after a controversial vote, he had the desired authorization to establish a new college with a separate faculty. In the new college, students could be admitted after two years of high school, and the curriculum consisted of a set of prescribed courses. For each course, there was a separate staff of faculty members. For students, what ultimately mattered was passing the examination for each course—it was possible to take the exams without attending the course. Passing all of the exams completed the B.A. degree. One unusually bright student, a future mathematician, passed all of the 15 general exams without taking the courses. As a result, he was able to start the serious study of mathematics at once.

The college initially consisted largely of inspired generalists; the staff for each course prepared the special materials for the course, making use of some of the Great Books. There resulted a heady

Chapter Twenty Nine — Hutchins and the University

college atmosphere, with an emphasis unusual for undergraduates, upon serious intellectual issues. Some of this came from the charismatic leadership of Hutchins himself—many students were vitally influenced and recall their studies as a climactic educational experience. Nearly all of the students were enthusiastic—rapturously so—about Hutchins and the college. Many faculty members also welcomed his vigorous defense of academic freedom.

Of course, this program was unusual and controversial, especially the ability to skip the last two years of high school. In some ways, it was really an early recognition of the weaknesses developing in the standard American secondary education system, but it did show that a university can provide a "high-voltage" intellectual experience for undergraduate students.

When I returned to Chicago in 1948, Hutchins was still president, although his title had been changed to chancellor, but he had withdrawn from some of the details of administration. At the start of his presidency, Hutchins, the boy wonder, had tended to manage many of the overall administrative tasks himself. Now, older and perhaps wiser, he delegated more. For instance, the administrative tasks associated with the Manhattan Project had brought several administrative hopefuls to Chicago; in particular, Lawrence Kimpton moved into administration there just after finishing his Ph.D. in philosophy. He was, in effect, the provost of the university at the time I received the Chicago offer in 1947, and I tried to negotiate some smaller issues with him without much success.

When Dorothy and I settled in Chicago in 1948, we invited Hutchins to dinner at our apartment, a dismal choice of an apartment located on the wrong (south) side of the Midway, and with only one bedroom for our two daughters. We enjoyed having Hutchins as our guest; at one point, I raised some administrative matter of interest to mathematics, and his vigorous response made it clear that he was still running things as he wished (though not as I might wish). But there is really a sense that a university president has a lonely life, since his relations with faculty members cannot be too close. I admired Hutchins, but found no effective way to say so.

Chicago in the Fifties

Several years after arriving at Chicago, I received an attractive (unsolicited) offer from another university. I did not really intend to leave, but I knew that the extraordinary new strength of the department and not my salary and perquisites were the real matter that might hold me at Chicago. So before deciding, I wrote Hutchins a letter describing the exciting accomplishments of our younger faculty members and the eventuality of needing to promote them to tenure. Hutchins responded in careful form: promotions might be possible, but they would need to be suitably justified at the appropriate time. I turned the outside offer down, having full confidence in Hutchins as well as confidence in my younger colleagues.

The university was under continuous and increasing pressures, some of which came from the state legislature, which, for example, became critical of an assistant professor of political science, Fred Schuman, whose political opinions were said to have corrupted one student, a godson of a legislator. Hutchins forcefully defended Schuman, though his book and ideas were not actually very radical. Chicago kept him on the faculty, but he did move away soon after receiving a better offer elsewhere.

The neighborhood of the university was also under pressure—many new people had moved to Chicago for wartime jobs. There were too many bars nearby on 55th Street, while the region south of the Midway was rapidly deteriorating. This did not cause our family any trouble, since Dorothy had providentially found an apartment (that she had long admired) at 5712 Dorchester Avenue. The apartment was a cooperative that had separate rooms for our daughters and close proximity to the Laboratory Schools, where we happily sent them for their education. But our good fortune was not universal—some faculty members moved to suburbs, while others advocated moving the whole university to William's Bay on Lake Geneva in southern Wisconsin, the location of the university's observatory. Many disapproved of the proposal, saying that "Astronomy can't be the center of the university," and "Wouldn't it be curious if the University of Chicago were not located in Chicago?" At this time, the University of Illinois was planning to start a more active branch in Chicago. Evidently, Hutchins decided not to sell the

Chapter Twenty Nine ~ Hutchins and the University

campus and buildings to the state, and obviously, he did not move the university.

But the bars on 55th Street were troublesome: One of my friends, a visiting philosopher, once had his friends wheel him toward a bar in a purloined baby carriage. The police stopped them, and a university dean had to negotiate with the police for his release. He was released, but shortly left for the University of Buffalo. As the German drinking song has it, *"Bier her, Bier her, oder ich fall um."* What should a president do?

I did not have any pipeline to the University Board of Trustees (admittedly a mistake; next time around I'll take care to join Skull and Bones or the next best equivalent to that secret Yale society), so I was taken by surprise in 1951 to hear that Hutchins had resigned. After 22 years in office, he was to become an officer of the Ford Foundation and then a head honcho of a think tank in California. When Hutchins left the university, the students held a three-hour reception for him filled with admiration. I wrote him a note of regret, and he responded with good wishes for the university. All told, he had wielded decisive influence as a trailblazing university president.

Lawrence Kimpton, the next president, took on the problems of the university's neighborhood as a major issue. At the time, there was an extensive government program for the renewal of inner cities, which was applied to Hyde Park. Many bars were closed, a big new apartment building was built on 55th Street (literally, in the middle of the street), and many new houses and row houses were constructed. As a result, the neighborhood dramatically improved.

The Hutchins College continued to function for a few years after Hutchins' departure, but gradually it came apart. On the one hand, the graduate (divisional) faculties did not like the college because they had no influence there and because its requirements kept students from early specialization. On the other hand, the continuing staff for each of the college courses grew weary of continually revising and teaching essentially the same materials. Soon, members of the divisional faculties were appointed to the college faculty, then, high school juniors were no longer admitted as college students. Perhaps the notion of a separate collegiate faculty in a research university is not really viable, since it does not provide for connection between research and teaching, so that ultimately, the teaching suffers. Or perhaps traditional Great Ideas hide the new great ideas coming from research. In any case, the presence of a separate college faculty and the great dream of general education were both gone.

Chapter Thirty — The College Mathematics Staff

As previously discussed, the Hutchins general-education plan at Chicago led to the creation of a separate faculty for the College. There was a mathematics staff, separately maintained from the Department of Mathematics, that focused on undergraduate teaching. The head of the math staff was my good friend and Yale classmate Eugene Northrop, while my former student, Alfred Putnam was a staff member. Herman Meyer, to whom I had taught algebra in 1937, was another member. The beginning course taught by the math staff was in favor of the Bourbaki approach to elementary mathematics with much emphasis on set theory. I disagreed with them on this point, but otherwise, relations went well.

On several occasions I searched for promising young mathematicians and recommended them for appointment to the math staff if there was no vacancy in the department. For example, I met Richard Lashof at Columbia when he was finishing his thesis there, and recommended him to Northrop; he soon joined the staff. While at Chicago, he collaborated with Eldon Dyer in topology to formulate the so-called Dyer-Lashof operations in topology, so good mathematics resulted.

After Hutchins left the presidency, there was no longer a permanent commitment to a separate college faculty. So, after time, the college mathematics staff was disbanded. Northrop became a representative of the Rockefeller Foundation in Turkey, where Dorothy and I later visited him. Some junior staff members left the university, while several others became members of the department; Lashof, for example, continued his research and served a term as chairman of Mathematics.

Chapter Thirty One
Universal Algebra and Think Tanks

I interrupt to cite a brilliant example of an unintended success at a California think tank that resulted in a decisive advance in category theory, reaching far beyond the initial uses of categories, which was to clarify the way algebra reflects topological structure.

F. William Lawvere, an Indiana farm boy, attended the University of Indiana at a time when Eilenberg (very briefly) and Clifford Truesdell (less briefly) were on the faculty there. From Eilenberg, he learned some basic facts about categories and probably discovered for himself the central notion of an adjoint functor (a notion later formulated explicitly by Daniel Kan while studying with Eilenberg at Columbia). Truesdell was an expert on classical mechanics, and the work of Leonard Euler; Lawvere saw some unfinished new ideas there. By the time Lawvere received his bachelor's degree, Eilenberg had already left for a position at Columbia, and Truesdell would soon leave for Johns Hopkins. Lawvere asked for Truesdell's advice on which university to choose for graduate study, and was told, "By all means, go work with Eilenberg!"

A semester later, I happened to be in New York on an invitation to visit the Rockefeller University, and I ended up spending much of my time at Columbia talking to Eilenberg and his associates. One day, Sammy told me he had a young student who claimed that he could do set theory without elements. It was hard to understand the idea, and he wondered if I could talk with the student.

At the time, almost all logicians held that the only firm foundation for mathematics was the Zermelo-Fraenkel axioms for sets, for which

Chicago in the Fifties

the basic primitive notion was $x \in Y$, meaning x is an element of the set Y. I had learned those axioms years ago when reporting on a paper by Zermelo for E. H. Moore's seminar. Despite my belief in sets, I undertook to listen to Lawvere. I listened hard, for over an hour. At the end, I said sadly, "Bill, this just won't work. You can't do sets without elements, sorry," and reported this result to Eilenberg. Lawvere's graduate fellowship at Columbia was not renewed, and he and his wife left for California, where he found a job in one of the many think tanks there.

Some time later, I was in Washington on some sort of business. When it was over, I climbed on a shuttle plane going to New York, planning to visit my mother in Norwalk. And on the plane, I saw Sammy. When I sat down beside him, he turned to ask me if I remembered Lawvere and his dislike for elements. And I did. Sammy reached into his briefcase and pulled out a fat document to say, "Here is Lawvere's thesis. You are the reader," and he went promptly to sleep.

After my visit with my mother, I returned to Chicago and started to read Lawvre's paper. By the time I was halfway through, I was persuaded that there was indeed exciting material there. It included a foundation for mathematics based on axioms (without elements) for the category of sets and functions. In addition, there was a wholly new way of doing universal algebra—instead of the customary start with a collection of n-ary operations for various n, there was a cleaner formulation using a category where objects were the natural numbers n, with object n the coproduct of n ones, and with the arrows $n \to 1$ all the n-ary operations involved in algebraic theory in question. In other words, Lawvere had used the basic idea of a category in ways going far beyond the original intentions; in particular, it included this strikingly different way of providing a foundation for mathematics. Neither Sammy nor I had contemplated any such direction. To this day, many logicians, and others, have failed to understand this approach, which has now been expressed using topos theory.

Later, I learned how it came to be that Lawvere's thesis was prepared at a think tank: When he first arrived, it seems that Lawvere worked with another young mathematician who also had an unfinished thesis. Lawvere talked with him about the problems in that thesis—

Chapter Thirty One — Universal Algebra and Think Tanks

they were quickly solved, and the other chap had his Ph.D. The think-tank boss was pleased, and, anxious to put his new Ph.D. to positive use, sent him off to one of those schools that train people for business jobs or, in other words, how to manage people. Upon returning to think-tanking, he was put in charge of several people, including Lawvere. He at once assigned Lawvere a new task: "Finish your thesis." That was how Sammy came to have Lawvere's thesis on that plane from Washington.

My own experience with think tanks, aside from wartime, is limited, thus I cannot generalize on their use. However, the model seems imperfect. For example, there had been some useful think-tank thoughts about operations research during the war, but when this was classified, as for example with the Rand Corporation, then its value was inevitably diminished. But I can observe that we should not conclude that the established way of doing foundations, or other parts of mathematics, is the only way.

Part Eight

Mathematical Developments

Chapter Thirty Two
Mathematical Organizations

Mathematicians in American colleges and universities have two major organizations. The first is the American Mathematical Society (AMS), founded long ago at Columbia. Its major concern has traditionally been the encouragement of research in mathematics through meetings and publications. As already mentioned, I joined the AMS early in my career to attend meetings and search for jobs. I also became active on various AMS committees, such as the Committee on Publicity, which tried without much success to encourage newspaper reports on the progress of mathematical research. Somewhat later, I became an editor, first for the Bulletin of the AMS, and later as the algebra editor for the *Transactions* of the AMS. In these capacities, I inevitably had to pick suitable referees for research articles and to decide which articles to accept, and I generally enjoyed the work.

The AMS emphasis on research left some gaps; for this reason, E. H. Moore and Herbert E. Slaught, both at Chicago, formed a separate organization to support and encourage the teaching of mathematics in college and education, the Mathematical Association of America (MAA). As its journal, it took over the already existing *American Mathematical Monthly*, which presented news, problems, and exposition of mathematical topics of interest to college teachers. The MAA also sponsored the *Carus Mathematical Monographs*, a series I edited for a period. For this reason, and for other activities in which I participated, I was elected president of the MAA for 1952–53. During this time, I encouraged the association to be more active;

the MAA secretary, Harry Gehman, who had been a Yale faculty member while I was a student there, helped me in these efforts. I viewed my election to the presidency of the MAA as an opportunity to keep the MAA and AMS together in spirit, since they served the overlapping interests of college and university mathematicians in research and teaching. There is no clean line of demarcation between these activities, and both organizations encourage clear exposition of mathematical ideas. On several occasions, I have even advocated combining the AMS and MAA into one organization, though I have not yet succeeded in this ambition. However, the Joint Mathematics Meetings of the two organizations are vital for the presentation and exchange of ideas, an exchange that is vital for both encouraging relevant research and effective teaching.

In addition to the AMS and MAA, there are other organizations for mathematicians, such as the Association for Symbolic Logic. Mathematical logic is a lively, but unusually specialized field of research whose practitioners can be in departments of philosophy or mathematics. The philosophical tradition, going back at least to the Greek philosophers, is perhaps the oldest, but logic became, and perhaps always was, necessary for mathematics. This need became explicit with the work of Peano, Russell, Whitehead, and others, but in the 1930s, there were many mathematicians who thought that logic was really not a proper part of mathematical research, and some mathematicians, such as Oystein Ore, who thought that logic did not belong at all in mathematics departments. Oswald Veblen took the first corrective step in his retiring address as president of the AMS, when he announced that logic could not adequately progress until more mathematicians took an interest in it. He then promptly acted on his words by encouraging the appointment of his student, Alonzo Church, to the Princeton faculty. Church proposed the so-called lambda calculus as a new foundation for mathematics, among other activities. Church's graduate students S. C. Kleene and J. B. Rosser, soon had a joint paper that proved that the use of lambda calculus as a foundation for all of mathematics was inconsistent. Happily, the lambda calculus has survived as an analysis of functionality and as a rubric for programming languages such as LISP.

Chapter Thirty Two ~ Mathematical Organizations

Logic stood in a special situation because it is sometimes studied by mathematicians, and sometimes by logicians, so eventually, in 1934, the Association for Symbolic Logic developed as a support organization, with its own journal, the *Journal of Symbolic Logic*. The journal had an auspicious start, with Church as its first editor, an annotated bibliography, and the practice of publishing the reviews of research papers in logic. Church served as an especially well-qualified editor; I, for one, enjoyed getting his knowledgeable comments on the reviews that I prepared. However, I do not mean to suggest that the Association or its journal was the dominant reason for the growth of activity in mathematical logic—the decisive steps were surely the famous Gödel incompleteness theorems and the discovery of the Turing machine. Mathematical logic, perhaps because of its isolation at the start, has continued to be a generally separate activity of mathematics departments, much more so than other specialties such as topology or algebraic geometry.

Still, these three organizations were not enough for the mathematics community—a Society for Industrial and Applied Mathematics (SIAM) was organized, with an impressive list of journals. More recently, there is also the Association for Women in Mathematics (AWM), and the National Association for Mathematics (NAM). There was, and is, still an evident need for an umbrella organization, which will be addressed in a later chapter.

Chapter Thirty Three
Bourbaki-the Legend

Legend has it that in the Franco-Prussian War of 1876, a French general, Nicholas Bourbaki, had apparently failed in battle. In a spirit of survival, his name was adopted by the students at the École Normale Supérieure as a rallying cry—when they wanted to riot, "Bourbaki" was their slogan, but now, at least to mathematicians, this slogan means much more.

By 1935, almost all of the prominent French mathematicians had studied at the École Normale, and in those days, French mathematics was dominated by the standard *Cours d'Analyse*, whether authored by Goursat or some other established mathematician. Those big, traditional volumes somehow did not come up to the more modern standards and objectives presented in German mathematics in Göttingen. Young French mathematicians teaching at Strasbourg—Weil, Cartan, and others—knew that something better and more modern in the way of *Cours d'Analyse* had to be prepared, so they set out to do this in the same revolutionary spirit as their students at École Normale, and they chose "Bourbaki" as their battle cry. First published in 1939, "Bourbaki" became a sequence of informative volumes that became legendary.

One version of the legend has it that this group of young and potentially rebellious mathematicians were walking in Montmartre by a café and observed an old *clochard*—a tattered tramp—sipping absinthe and muttering to himself "compact spaces, differential forms…" So they sat at his feet and learned how modern mathematics should be written. And write it they did. Somewhere, I have a handsome picture of the old clochard, beard and all, as well as copies of the resulting, informative volumes. Whatever the old man's influence, the Bourbaki volumes organized and systematized a great deal of mathematics, but to my regret, they neglected categories. Henceforth, Bourbaki and his ideas were noted wherever mathematicians gathered.

In about 1950, Ralph Boas, a mathematician at Northwestern University, wrote the article about mathematics in the Encyclopaedia Britannica's annual volume of events of the year. For the year, he reported that Nicholas Bourbaki had published another volume of his impressive treatise, and that—as everyone knew—Bourbaki was simply a pen name for an officially anonymous group of French mathematicians.

Weil happened to read the article, and at once wrote the editor of Britannica in the name of Bourbaki, vigorously asserting that he did indeed exist, and that, in fact, had recently given a public lecture by invitation of the ASL. He added that the editor could check on this matter by consulting Bourbaki's friend Saunders Mac Lane at the University of Chicago. Weil then came to my office to drop a copy of the letter on to my desk, warning me, "Saunders, if you don't tell him the truth, I will never speak to you again." Wow!

It was the case that Nicholas Bourbaki, in the person of André Weil, had recently lectured on the foundations of mathematics at a meeting of the ASL, but a recounting of that spin of the truth to the editor would have served no purpose. I wrote an ambiguous letter to the editor; fortunately, Weil did not stop speaking to me. However, Bourbaki himself did perform the effective action of spreading a rumor that Ralph Boas did not exist.

The Bourbaki volumes did represent an unusual example of selfless collaboration, and the only thing Bourbaki never understood was the foundation of mathematics: Since foundations should come first, that volume was written early on, resulting in their very clumsy description of a mathematical structure and its resulting clumsy attempt to formulate an existing theorem for free objects with a given structure. Peter Freyd's later proof of the adjoint functor theorem, which gave the existence of adjoints, really clarified this situation, since free objects are an example of adjoints.

Chapter Thirty Three ~ Bourbaki-the Legend

Bourbaki's foundations volume was written before the ideas of category theory were generally known. In 1954, I was invited to attend one of the private meetings of Bourbaki, perhaps with the prospect that I could help to introduce some category theory there. I found the meeting fascinating, with its lively debates and occasional excursions into the country, but my fluency in French was not sufficient for me to take a persuasive part in Bourbaki's discussions, and the spirit of the organization was clearly and distinctly French.

Foundations aside, the Bourbaki volumes present a wide-ranging and inspiring organization of mathematics, carried out by a talented and selfless group of mathematicians who have had a major influence on mathematicians everywhere.

Chapter Thirty Four
The New Math

By 1940, the many natural connections between mathematics as taught in secondary schools and mathematics in colleges and universities had become very weak. High-school teachers were simply teachers, expected just to train their students to manipulate numbers, manage algebra and perhaps even to understand that Euclidean geometry rests on axioms. About 1955, my own connection to the International Mathematical Union brought me to prepare a sort of report on secondary school mathematics education in the United States—the actual title was "Algebra," as published in the *Yearbook of the National Council of Teachers of Mathematics*. It was my opinion that new ideas should be brought in to secondary schools, and teachers should be better trained.

And then politics intervened: when the Soviet Union succeeded in launching Sputnik in 1957, a good deal of American complacency vanished. The Russians, it seemed, were ahead of us, and we needed more high-school graduates who really understood mathematics as a basis for other sciences. The result was a crash program called new math, which was a concerted effort to improve grade-school and high-school mathematics.

The new mathematics was to be more precise, and it was clear and fortunate that Bourbaki had set forth the high road to precision: use set theory and define functions as sets of ordered pairs. The emphasis on precision was widespread, and was held particularly firm in the College Mathematics Staff at Chicago—my friends tell me that I argued against this, but without success. This view was by no means unique to Chicago—in some places it reached down as far as kindergarten, where lucky students might be taught that a set consisting of exactly the two elements a and b could be symbolized as $\{a, b\}$. And the expected story ensued: Johnny's parents ask the kindergarten teacher how he is doing, and she tells them "Yes, he understands sets, but he has trouble writing the curly brackets."

Even before Sputnik, it was clear that teachers of mathematics need much more training and more knowledge of real mathematics. To this end, the National Science Foundation supported teacher institutes at many locations; some were just summer institutes, others were yearlong. It is my impression that they worked well. I took part in one Academic Year Institute from 1957–58 at Chicago, in which 40 high-school teachers and prospective teachers spent a year at the university learning about how to look at mathematics and how to better teach it; the teachers attending were by and large enthusiastic.

I became interested in the way geometry was taught, and in this connection, came across the Birkhoff-Beatley geometry. At Harvard some years before, George Birkhoff became interested in school geometry—he found the standard Euclidean axioms too elaborate and cumbersome, and developed an alternative that first assumes all the standard properties of the real numbers as axioms. Then, geometric objects can be described rigorously in terms of pairs of coordinates. Birkhoff and his colleague, Ralph Beatley, systematically developed such results in a high-school textbook. I had known essentially nothing about this approach while at Harvard, and had fun exploring it during the year I spent with the teacher institute at Chicago. The traditional Euclidean geometry has merits, too; for example, it teaches proof, with the usual two-column proof format, but I thought it was good to explore new ideas.

The institutes received great enthusiasm, and they went on around the country for several years in the late '50s and early '60s. In 1958, however, President Kimpton decided to establish a school of education, which meant, it seemed to me, that any future institutes would be under the control of the school of education, not the math department. I disagreed with this approach, and when it was being considered by the Faculty Council and Faculty Senate, wrote a letter

Chapter Thirty Four ⁕ The New Math

to Kimpton strongly expressing my opinion. It seemed to me that the effort in the mathematics department had been one of the major attempts at the university to do something about the situation in secondary education, and I wrote the letter because I thought Kimpton was ignoring the real activity that was going on. The school of education was nevertheless established, and I was sufficiently fed up that I didn't do much more in secondary education in the next year.

The teacher institutes for new math successfully brought active high-school teachers into a collegiate environment where they could learn new ideas about what was really going on in mathematics and meet colleagues, as well. In our Institute, new math was much more than just the expression of precision in set theory; it was a consortium of progressive high-school teachers and university faculty working together to improve mathematics education. New math is often remembered more for its extreme views, but it also had substantial results; for example, the introduction of courses in calculus and statistics at secondary schools. However, the teacher institutes paid little attention to traditional pedagogical ideas, and the NSF funding was eventually canceled; a program director at the NSF told me that the Department of Education encouraged the cancelation.

Chapter Thirty Five
Categories Expand

In 1945, when Sammy and I succeeded in publishing our systematic paper on categories, "The General Theory of Natural Equivalences," Sammy announced that this would be the only paper necessary on the subject. It turned out that he was wrong. Probably, he had just the first use of categories in axiomatized homology theory in mind, but later, more general concepts about categories were developed and applied.

One of the first additional notions to appear was the abelian category, like the category of all abelian groups or that of all R-modules over a given ring R. Here, the essential added element of structure is the assumption that any two morphisms $f, g: A \to B$ between the same two objects can be added so as to make the set $Hom(A, B)$ of all such morphisms into an abelian group. Such abelian categories, suitably axiomatized, appear in algebraic topology, where the homology and cohomology groups on the category of topological spaces take values in an abelian group, that is, an abelian category. This application is the one I had in mind when I first introduced the notion in Zurich in 1948: the axioms I presented were somewhat clumsy and were later improved by Buchsbaum in his 1955 Columbia thesis. At first, the idea of abelian categories was not widely adopted. A few years later, Alexander Grothendieck, who had been working on functional analysis, came to Chicago on Weil's invitation to give a lecture in which he used abelian categories as one of his tools. I did point out to him that the notion was previously known, with reference to my 1948 paper; however, it is entirely possible that he had discovered and named abelian categories on his own, but it is also possible that he had learned about them from his frequent contacts with J. P. Serre in France. It was after this that Grothendieck published his famous paper "On Homological Algebra" in the *Tohoku Math Journal*, which set the stage for his extensive use of categories in his subsequent, influential reformulation of algebraic geometry. Originally suggested by problems in algebraic topology, categories, by now, had found this new extensive use in evidently related aspects of algebraic geometry; they form a remarkable example of the underlying unity of mathematics, as expressed in the use of common general concepts in widely different fields of study.

Eilenberg and I were fortunate to have been there at the start—I have sometimes speculated as to whether someone else might have discovered category theory had we not been there at the start, and have found no convincing answer to such speculation; the very generality of the notion would easily have been quite intimidating.

The next general categorical idea to develop is that of a universal construction. Given a group G and a normal subgroup N, the familiar factor group G/N is the image of a homomorphism $G \to G/N$ that sends all the elements of N, and only these elements, to the identity. Any other homomorphism $G \to H$ sending N to 1 must necessarily factor through the map $G \to G/N$ as $G \to G/N \to H$. In other words, the factor group, often described as a group whose elements are the cosets Ng, can instead be described more conceptually as the solution of a universal problem: find the essentially unique homomorphism from G in which all the elements of N are sent to the identity. In this sense, the use of cosets as the elements of G/N is unnecessary. What really matters is the universal properties of the factor group. This is an idea that quite evidently would have pleased Emmy Noether, but it is, as yet, not sufficiently used in modern algebra courses.

Chapter Thirty Six
The Grand Tour of Europe, 1954

In 1954, with the prospect of an International Congress of Mathematicians in Amsterdam, Dorothy and I decided to give Gretchen, then 16, and Cynthia, then 13, the traditional grand tour of Europe. We traveled chiefly by car, and I can recount the journey through a letter we wrote to my mother titled "Through Europe with Four Wheels, Three Girls, Two Appetites, and One Camera."

As was then customary, we started by ship, the Dutch Maasdam, sailing from New York on June 14, 1954. Describing the beginning of our voyage I wrote, "The Dutch are very efficient, but do not mind inconveniencing the passengers for the sake of this efficiency. Immediately after getting to the cabin, I had to fill out a declaration of everything about our destination. Next there came the dining-room reservations. As soon as we got on the boat, Dorothy went up to stand in line in the lounge for table reservations. She was still in line when the boat was ready to leave the pier, so I went up to relieve her so she could see us depart. By noon, I was at the head of the line and finally managed to get four places at a big table for the second sitting, so we didn't have to get down for breakfast until 9:00 a.m.

Cynthia and Gretchen immediately took to life on the boat. They were inducted early into deck games such as deck tennis and shuffleboard. We got four deck chairs on what was probably the windiest side of the boat, but Dorothy seemed to like it that way—she sat out there reading books from the ship's library. Our dining room table presented a considerable variety of people: an Irish blacksmith from Canada who was going back to Europe for the first time in 35 years; an elderly lady who was once a refugee from Germany; an opera singer; a middle-aged lady; and two young men. Fortunately, Gretchen sat near the young men and I was left to listen to the blacksmith's life story. A steward announced meals by walking around and playing a tune on a small xylophone. A couple of days before the end of the trip we heard the usual tune and looked up to find Cynthia playing the xylophone. The steward was only 19, and all the girls thought he was cute.

We arrived in Southampton by way of the Solent, which, I learned, is a channel between the mainland and the Isle of Wight. We went first to Winchester and saw the famous cathedral that is said to be the largest in England. We also took a side trip to Portsmouth so that I could see the HMS Victory, Nelson's flagship at Trafalgar, which is propped up in dry dock and under almost continuous repair.

Cynthia and Gretchen both had dramatic reactions to being in foreign parts. We had not told them that many things would be different abroad, even in England where the language was the same as our own. They discovered this the hard way; for example, one afternoon they got thirsty and wanted something cool to drink—they were so anxious that we stopped at the first refreshment stand with a Coca-Cola sign. When they asked for a coke, they had to explain that they meant Coca-Cola; they finally got it, but it was warm. At 6:00 in the evening, they were hungry, and we could not easily find a restaurant. They were disgusted with the observation that there were none at hand, and insisted that some must be open, even if the British did not eat till 7:30. We looked at cafes, but could not understand that they really were not restaurants. Finally, we had tea at King Alfred's Refreshment Stand. Gretchen discovered, to her horror, that the tea was half milk, and she finally understood why King Alfred had such a hard time of it. Disgusted with British tea, she insisted on having coffee for breakfast, but then discovered that the coffee was black and strong, and she turned her nose up at that too.

From Winchester, we went by bus to Oxford, where we sent the girls off by themselves for a boat trip on the Thames. Dorothy and I

Chapter Thirty Six — The Grand Tour of Europe, 1954

had dinner with my friend, topologist J.H.C. Whitehead, who was living in the country on a centuries-old farm—he did mathematics while his wife did the farming. We stayed at the Golden Cross, which was very old, and it boasted that Shakespeare stayed there. The next day, we explored the colleges, Lincoln and Brasenose, and were about to go into a third when Gretchen demurred, saying that they were all alike, at best. That evening, the girls had dinner by themselves, Dorothy had dinner with Mrs. Whitehead, and I had dinner with Henry and two of his students, Ioan James and Victor Guggenheim.

The next day, we went by bus through the Cotswold, Chipping Norton, Stier-on-Stow, and other places with bizarre names; Dorothy was sure they were significant, but did not know why. When we arrived at Stratford, we found that the Avon was peaceful. We thought we had reservations at the Royal Hotel, but they claimed we had not confirmed them—despite this, they found rooms for us. We inspected Shakespeare's birthplace, which was thronged with people; had high tea (to replace supper); and saw *Othello*. After the play we could find no restaurant for supper, and so went to bed hungry.

After that, we visited the Lake Country, then on to Glasgow and Inverary, and then by boat to Tobermory on the Isle of Mull. We had called up Sir Charles Hector Fitzroy MacLean, who encouraged us to visit Duart Castle, the ancestral home of the clan MacLean. The high and foreboding castle is situated on a bluff at the edge of Mull Sound, which was in ruins from 1745 (at the time of the Bonnie Prince Charlie) until 1911, but has now been fully restored. In spite of its eight-foot thick walls, it is very livable, with a mammoth banquet hall including a fireplace and ancestral pictures, and a master bedroom with a magnificent beamed ceiling. Even Cynthia, who had previously labeled this part of our expedition "ancestor worship in modern style," enjoyed the castle. From the Isle of Mull, we went by excursion boat to Staffa and Iona, the latter the sacred island of Scotland where Saint Columbia from Ireland had established a Christian mission in the eighth century A.D.

Next, we were off to Inverness and then Edinburgh, which is really two towns; the old town grew up around the castle on top of a steep hill. From the castle, the Royal Mile leads downhill to Holyrood

Mathematical Developments

Cynthia and Gretchen feeding pigeons in London, July 10, 1954

Palace, where we saw the small room once inhabited by Mary, Queen of Scots, with the secret stairway to Darnley's room, and the spot where Mary's secretary and confidant, Rizzio, was murdered.

From Edinburgh we went by night train to London. There, we saw St. Paul's enormous baroque cathedral, Trafalgar Square with its pigeons to be fed, the Tate Gallery with many Turners (too many), the Tower of London, and *The Mousetrap* by Agatha Christie—all good.

The guard does change at Buckingham Palace, although only every other day, except during tourist season. We got there and took three rolls of film—lots of other people took more. I conjectured that the whole ceremony is just for the benefit of the camera-film industry.

We next went by train to Harwich, and then by overnight boat to Rotterdam. There, we rented a car, a 1953 Opel Record Olympia, for the rest of our tour. The car was convenient, but slow at all border crossings. I had a very thick wad of car papers—at every border crossing they took off a paper, just to ensure that I was really not importing the car to sell it.

From Rotterdam we went to Utrecht and then to Arnheim to cross the border between Zevenaar (Dutch) and Emmerich (German). Then on to the university town of Münster, where I gave a lecture to 200 students in the morning and a technical lecture in the afternoon. We then drove to Göttingen, and the town was just as I remembered it, but with everything more orderly and prosperous than at my

Chapter Thirty Six — The Grand Tour of Europe, 1954

previous visit in 1948. Professor Franz Rellich and several other German mathematicians entertained us well at dinner; most of the conversation was in German, to the considerable annoyance of Cynthia and Gretchen. Before this time, they had fondly believed that foreign languages were mostly an invention of schoolmasters, intended to torture and annoy children. Only here did they realize that there were actual people who talked in these strange ways, and either could not, or would not, speak English.

We took a walk all the way around the town on the top of the old town wall. Then, we went by autobahn through Hanaversal-Munden and Hamburg, and other lovely hill-and-castle towns. We were then on to Marburg, where we met Professor Reidemeister, a well-known topologist, and Professor Arnold Schmidt, a logician I had known when we were students at Göttingen. We then went to see my friend, Professor Franz, in Frankfurt am Main. The problem was that the reservations were not under Mac Lane or Saunders, but simply under Lane, but they were there and they were comfortable.

Much of Frankfurt had been rebuilt since my former visit in 1948; narrow streets had been widened and new squares created in the city center. The house where Goethe was born had been precisely reconstructed—the new floorboards were exactly the same width as the originals and the nails were even in the same places. The first evening, I lectured on mathematics; in the first five minutes, I managed to lose the audience, but after hard work I got them back again, which is unusual.

Cynthia and Gretchen much enjoyed seeing the romantic Heidelberg castle. Next was Rothenberg ob der Tauber and Dinkelsbühl, an old walled town with impressive, half-timbered houses. We continued to another university town, Tübingen, also on the river Neckar. There, I again saw friends (Professors Knopp and Wielandt) and gave a lecture. After the lecture we had a dinner with many mathematicians. This time, we managed to seat our daughters next to young, unmarried, English-speaking mathematicians, but alas, afterwards, the girls announced that they were not cute.

As we drove on, we passed the well-marked watershed between the Danube and the Rhine; I took a picture of Dorothy by the stone

Mathematical Developments

Dorothy by the stone marker of the Danube-Rhine watershed, 1954

monument to the divide. Then, we drove on over the border to Zürich, where both girls bought new watches and swam in Lake Zürich.

From there, we aimed for Italy, and crossed through the Alps by the Gotthard pass; this took us up through the lovely town of Zug and to the Lake of the Four Cantons, where we had tea on the terrace of a splendid hotel overlooking the lake. But going up the Gotthard is quite a climb—halfway up the steepest parts, we came across a one-way traffic section where we had to wait on the side of the mountain, leaving the motor running as instructed. The motor coughed and sputtered, and the daughters feared that it would die but, fortunately, it did not. On top of the pass we had a considerable view, and we then started down in low gear, negotiating a dozen or more switchbacks. I was glad I didn't have a Cadillac—its length would have been a bother on those many turns. We got to Lucerne at about 7 p.m., and, luckily found a hotel, since we had made no reservations. From the hotel, we rode a funicular railway halfway up the mountain to get a splendid view of the city below us at night. In the morning, we had a swim in Lake Lucerne.

Lugano, another tourist town, was the next stop, and we then crossed into Italy, where we visited Florence, Venice, Milan, and Naples. The border crossing took time because we had to negotiate about the discount coupon for gas supposedly available to tourists. We had trouble because our car had been in Italy once before in the year.

Chapter Thirty Six — The Grand Tour of Europe, 1954

Italy was really different. The first thing we noticed was the driving, which was done with the horn: When in doubt—honk; if you come to a curve, don't slow down—honk; if danger threatens, don't brake—honk; if you want to pass someone—honk. At first, my execution of these basic ideas was pretty poor, but Gretchen and Cynthia rapidly noticed what was going on and bravely took on the task of coaching me in the use of the horn. Gradually, I became better, but I never did bring myself to cut blind curves with a honk.

After the grand tour, we attended the International Conference in Amsterdam and finally, returned home on the Maasdam.

Chapter Thirty Seven
Paris and Cartan, 1955–56

One of the major attractions in the mathematical world of the 1950s was the seminars that Henri Cartan conducted in Paris, which dealt with current topics and analyzed most recent developments, often expanding or simplifying results that had been obtained very recently. The lectures given at the seminar were published as mimeographed notes and provided a much sought-after source of inspiration to young mathematicians the world over. In 1954–55, the topic was Algebras of Eilenberg-Mac Lane and Homotopy, which concerned our work on the spaces $K(\pi, n)$, those spaces that have only one nonvanishing homotopy group, namely the group π in dimension n. We had described a bar resolution B that would construct a model (a differential algebra) for $K(\pi, n+1)$, given such a model for $K(\pi, n)$. I decided to spend the following academic year in Paris, and with the help of a Guggenheim Fellowship I took a leave of absence from Chicago and moved to Paris with Dorothy and the girls. Adrian Albert took over my position as chairman in my absence.

The chief reason for heading to Paris was the lure of the Cartan seminar. In the published notes of the previous seminar, the participants analyzed the sorts of constructions that were involved in our approach and gave a more conceptual way of passing from $K(\pi, n)$ to $K(\pi, n+1)$ that I found stimulating and insightful, and I still think that these methods deserve further analysis. During the year in Paris I did not manage to do this, though two of my students,

Mathematical Developments

Gretchen and Cynthia at the apartment in Paris, December 1955

Gerald Decker and Ross Hamsher, worked in this direction and published their work.

For my family, the stay in Paris was exciting and stimulating. We rented a roomy apartment at 72 Avenue de la Bourdonnais, a pleasant setting near the Invalides. The bedrooms were adequate and the living room fine, and there was also an inside winding stairway that led up to a locked door that belonged to a bedroom once attached to the apartment. However, nothing is ever perfect and my daughters remember problems with the apartment; it seems that my decision to rent was based partly on the inclusion of a refrigerator in the rental. Dorothy had specifically asked that it come with the apartment, but when we arrived, the refrigerator was far less than adequate—it held about two bottles and dated from the 1920s. Dorothy wailed the moment we walked in; she would be, then, as she explained, a prisoner of the French shopping system, which involved shopping three times a day—once for each meal. We tried to live with it, but that was not possible for an American family of four, especially since Dorothy believed children should drink a quart of milk a day. Within a week or two, I bought a new, adequate, French refrigerator. However, Dorothy was still a slave to the French grocery system—they didn't have the frozen vegetables and so forth that she made for dinner, and

Chapter Thirty Seven ⁓ Paris and Cartan, 1955–56

she still had to shop twice a day for the bread, among other things. It may seem unimportant, but such little cultural differences can spoil an otherwise wonderful experience. In spite of this beginning, Dorothy enjoyed the city and the sights, and occasionally entertained guests after settling in. At one point, she put on a party and ordered a quart of whipped cream; somehow, the order was changed to a gallon, so we had much extra cream for a while. But the party went well, aided by the refrigerator.

On a number of occasions, Marc Krasner invited us to dinner at his apartment—he loved to give dinner parties for the mathematics department of the Sorbonne. Gretchen remembers the night we returned from the first party: We got home around 11, and Dorothy immediately commented on Krasner's many cats, who were allowed free range of the apartment at all times. Thus, the cats were on the table alongside the dinner guests competing for the hundreds of oysters Krasner served as the first course. Dorothy loved oysters and hated cats with an equal passion; she shuddered in disgust about a childhood memory of an encounter with a kitten—she couldn't bear cats in any way and did not think they were at all cute. But oysters! Together, we once devoured nearly three dozen that Gretchen brought for lunch on a Cape Cod visit. Dorothy talked about Krasner's party for days—after that, he was persuaded to keep the cats in his bedroom while the guests were there. But at the next dinner, with Mr. and Mrs. Cartan, Mr. and Mrs. Dubreil, and other prominent members of the department in attendance, the cats managed to cause trouble again. Although he had shut them away in the bathroom, one cat somehow escaped. It had been a splendid many-course dinner prepared by Krasner's housekeeper, but the dinner went so slowly that some guests started to leave before the cognac, much to Krasner's distress. When Madame Cartan came to leave (just in time to catch the last Metro), she discovered, to her horror, that her hat was ruined—it had been slept on and squashed into a hairy pancake by the cat.

Gretchen, who was 17, delighted in Paris, in part because it was a center of dance. I remember her taking daily dance classes. She studied with Olga Preobrajenskaya, who had danced with Nijinksy at the Maryinsky in St. Petersburg. These classes continued Gretchen's interest in the art of dance, and I am delighted that our visit had good outcomes beyond mathematics. Cynthia, who was 13, entered as a student in the École Alsacienne, which we knew to be a school suited to students from Alsace, and even more for students from foreign countries. When I first went with Cynthia to register her, she was astounded by my apparent mastery of French, but after three months of schooling she spoke fluent French, and constantly corrected my inept pronunciation of that beautiful language.

Despite my uncertain mastery of the language, I managed to give several lectures, and I finally managed to carry through my earlier ideas about the homology of rings. The leading idea is that the two-dimensional cohomology $H^2(R, A)$ of a ring R with coefficients in an abelian group A (regarded as a ring with all products zero) should describe all the ring extensions of A by R; that is, all short exact sequences of maps

$$0 \to A \to S \to R \to 0$$

of the ring homomorphisms. For each x in R choose $u(x)$ in S; then, the addition and multiplication in S is determined by two factor sets f and g presented by the equations

$$u(x) + u(y) = f(x, y) + u(x + y),$$
$$u(x)u(y) = g(x, y) + u(xy)$$

for addition and multiplication, respectively. These factor sets satisfy various identities that reflect the laws holding in the ring S. They lead to a two-dimensional cohomology group $H^2(R, A)$. What I did was construct a cohomology theory $H^n(R, A)$ extending this two-dimensional theory to all dimensions n. This is a refinement of the standard Hochschild cohomology, which in dimension 2, concerns only the factor sets $g(x, y)$ for multiplication alone. While in Paris I developed this theory and presented it in a conference at Louvain. U. Shukla has subsequently studied this idea and it is now called Shukla cohomology; as always, it is a pleasure to see how new ideas spread.

Although I remember the year as a productive one for me, the circumstances of the cultural change, both domestic and scientific, contributed to a temporary unhappiness that my daughters seem to

Chapter Thirty Seven ~ Paris and Cartan, 1955–56

remember more than I do. I might have felt frustrated because I arrived too late to be part of the exciting progress made in Cartan's seminar during the previous year when it had focused on my work. In the beginning of the year in Paris, I did not make progress in my work, but by the end of the year, I did have success in my efforts in homology theory to take home with me. Research has its ups and downs—the triumphs are mixed with disappointments.

When the summer came, we sublet the apartment to my friend and Chicago colleague, Irving Segal, and his wife Osa while we vacationed on Terschelling, one of the Dutch islands. After our year in Paris and vacation in the North Sea, we returned, refreshed, to Chicago—weather and mathematics alike.

Part Nine

National Academy of Science

Chapter Thirty Eight
Membership in the National Academy of Sciences

In April 1949, shortly before my 40th birthday, I was elected to membership in the National Academy of Sciences. This was an early election—Eilenberg was elected in 1959, and Garrett Birkhoff in 1968. The early election turned out to have consequences for me, leading to my extensive activity in Academy affairs, and eventually, to my service as vice president for two terms from 1973–1981. In the same year, I was also elected to membership in the American Philosophical Society, which was founded by Benjamin Franklin for "promoting useful knowledge" and is headquartered in Philadelphia. Earlier, I had become a member of the Boston-based American Academy of Arts and Sciences.

I attended my first meeting of the National Academy in the spring of 1950, and it turned out to be a dramatic occasion: Dr. Alfred N. Richards, who was then president, had decided to resign his office. As was customary, the nominating committee considered the matter and nominated James Conant, president of Harvard University. As soon as I arrived at the meeting, Wendell Latimer, a chemist from Berkeley, buttonholed me to ask what I thought of Conant as president at Harvard—it appeared that a number of chemists and other Academy members had found that Conant had been autocratic in some of his actions as vice chairman of the OSRD (the Office of Scientific Research and Development) during the war. Also, many Academy members felt that the Academy needed to be more actively represented on the scene in Washington. They admired the work of Detlev Bronk, who was then chairman of the National Research

Council (NRC) and also president of Johns Hopkins University. He was visibly active in both capacities. Latimer, among others, felt that if Conant became president of the Academy, he would incline to be an absentee president, spending most of his time as president of Harvard.

With these matters in mind, Latimer and others planned to present Bronk as a second candidate for president at the upcoming business meeting, which was quite an unusual action. There was considerable discussion in the corridors. One of Conant's friends, hearing of the situation, telephoned Conant, who withdrew his name. Bronk was thereupon elected president of the Academy.

This was a definitive result: The members wanted a more-active academy that was effective in pressing the needs of scientific research, and that is what they got. Bronk was indeed active, and he was succeeded by other active presidents, such as Frederick Seitz, Philip Handler, Frank Press, and Bruce Alberts. This sort of activity came at a suitable time, when the continued support for scientific research by the federal government played a growing role. The Academy played a decisive part in representing the needs of science to the government in Washington.

Chapter Thirty Nine
The National Research Council

The National Academy of Sciences was founded in 1863, during the Civil War. Its charter, signed by Abraham Lincoln, specifies that the Academy shall, "whenever called upon by any department of the government, investigate, examine, and report upon any subject of science or art" (in this context, "art" in effect meant engineering). But after 1865, the Academy members paid little attention to this prospective advisory function—they were more concerned with the honor of election to Academy membership. Then, during World War I, various military developments required significant, immediate scientific advice and help; therefore, the National Research Council was founded for this immediate purpose.

At the end of the war, President Woodrow Wilson instructed the Academy to "preserve and continue" this council, so from 1918 to 1940, the NRC served under the auspices of the NAS as a sort of holding company for the political interests and activities of the various American scientific societies. Generally, these societies had some type of official part in the governance of the NRC. The NRC was not a very active organization, but it was known for its stewardship of the National Research Council Fellowships, which were awarded competitively to beginning research scientists, and a considerable number of gifted young mathematicians then held such fellowships.

Detlev Bronk had been chairman of the NRC before becoming president of the Academy in 1950. Once president, he used the NRC as the working arm of the Academy for advice to the government— the reports the Academy issued to the government then were issued as the reports of the NRC. In this way, Bronk finally gave real substance to President Wilson's instruction in 1918, and the Academy control of the NRC led to the development of a Report Review Committee, a committee for which I later served as chairman.

Unlike Bronk, the members of the Academy were most interested in the honor of Academy membership, and hence, in the election of new members. Election is a careful process, requiring nomination by a section of the Academy or by a voluntary nominating group, written ballot of Academy members, and finally election at the annual spring meeting. The present annual quota for election is 60. In earlier times, the qualifications of individual candidates were discussed at the spring meeting. This general discussion no longer occurs, but the bylaws provide the option of an open challenge for individual candidates, which is an effective means of upholding quality.

Membership of the Academy is divided into sections according to scientific interests; each new member chooses a section, and each section is naturally anxious to improve its overall representation. In this process, the engineers seemed to have gradually lost out; for example, Thomas Edison was not elected to membership in the Academy until 1927 when he was 80 years old. There were a number of other cases where the interests of engineers were neglected. As a result, there was a proposal to establish a separate Academy of Engineering in 1963 under a new federal charter. This proposal troubled the officers of the Academy of Sciences—they feared that it would lead to divisive competition in the work of the NRC in advising the government. Frederick Seitz, who was then president of the Academy, proposed the establishment of a National Academy of Engineers under the existing charter of the National Academy of Sciences. This proposal was accepted, with the provision that the new academy would elect its own members, and the new academy would also share in the guidance of the work of the NRC.

There was considerable friction between the two academies at the beginning, but it died down in time. However, it did set a precedent—the doctors and the medical people thought that they should also have their own academy under the NAS charter. Instead of a separate academy, an Institute of Medicine (IOM) was developed in 1970,

Chapter Thirty Nine ~ The National Research Council

with a thoughtful proviso that qualified people willing to work on advisory committees should be elected. From that time on, the work of the NRC has been supervised by all three of the organizations—the NAS, the NAE, and the IOM, all through representation on the Governing Board of the NRC. Despite the inevitable, though occasional, conflicts, this quadrupedal organization has continued to function, producing a considerable volume of reports to the government. These reports are subjected to quality checks by the Report Review Committee, which I will discuss in detail later in connection with my work as chairman of the committee.

Chapter Forty
The Academy Proceedings

In 1913, the National Academy of Sciences established its journal, the *Proceedings of the National Academy of Sciences* (the PNAS). The journal was intended primarily to publish brief articles by Academy members, or articles by other scientists communicated by an Academy member. There was an editorial board made up of Academy members, and a managing editor. Edwin Wilson, who served as the first managing editor, had been a student of mathematics with Josiah Willard Gibbs at Yale, with whom he had written a widely used text on vector analysis. From Yale, Wilson moved to MIT, where he wrote a text on advanced calculus. He worked in many branches of mathematics, notably in mechanics and aeronautics, and later in statistics. In 1922, he shifted to the Harvard School of Public Health and took up statistics as his major activity. Wilson was the managing editor of the *Proceedings* for 50 years and wrote a fascinating volume on its history.

Up until 1947 or so, the *Proceedings* published many papers on mathematics, including many brief announcements of new results that would normally be published in full later in regular mathematics journals. Eilenberg and I published announcements of our work on categories, and later on cohomology groups. Rumor has it that biochemists used the same strategy; one young biochemist, whose paper was rejected by the journal of *Biochemistry*, managed to persuade a member of the Academy to publish his results in the *Proceedings*. This paper later led to a Nobel Prize. It may be that Wendell Stanley, who was then chairman of the editorial board, used this result to advertise the benefits of quick publication of biochemistry results in the *Proceedings*. For whatever reason, there came to be many such biochemistry papers.

In 1958 I was elected a member of the Council of the Academy, the Academy's governing body. In 1959, just before a meeting of the Council, the Academy president, Bronk, received news that Stanley was ill and wished to resign his position as chairman of the editorial board. The Council met, and Bronk liked to act promptly—he was not a man to put off until tomorrow what might be done today. He knew that the *Proceedings* carried a lot of math, so he looked about the table in that splendid boardroom, spotted the only mathematician there, and proposed to the Council that I be made chairman. Most likely, they did not know that I had been on the editorial board of the *Transactions*, which was the flagship journal of the American Mathematical Society. They probably appointed me because then, as now, the Council did not often disagree with the president.

Soon after my appointment, the treasurer of the Academy, concerned about costs, requested the introduction of page charges for papers published in the *Proceedings*. At that time, mathematicians did not have large grants, so most of them stopped submitting to the *Proceedings*. As a result, my editorial responsibilities did not include many mathematics papers, but I was pleased to work with Wilson, whose advanced calculus book I had once studied. In his capacity as managing editor, he took care not to set policies for the *Proceedings*, but he was happy to give me advice. In the course of about four years, he wrote me some 50 letters, mostly about the journal, but also about past members of the Academy.

I served as chairman of the editorial board for eight years. During this period, I used no referee system: papers by Academy members and papers communicated by them were almost always accepted. I recall a couple of exceptions, including one case in which I rejected a paper dealing with Vitamin C as a cure for cancer that had been communicated by Linus Pauling, but I no longer remember the details. In 1967, I resigned my position, and John Edsell, a biochemist, took over as chairman. A subsequent chairman has introduced a system of refereeing papers, even those written by academy members that has resulted in some unjustified rejections.

Part Ten

The Sixties and Beyond

Chapter Forty One
Homological Algebra

The cohomology of groups, as already described, had originated in the observation that the fundamental group π_1 of an aspherical space would determine all the homology and cohomology groups of that space—the latter as the cohomology $H^n(\pi_1, -)$ of the group π_1. From this result developed the surprising idea that cohomology, originally studied just for spaces, could also apply to algebraic objects such as groups and rings. Given his topological background and enthusiasm, Eilenberg was perhaps the first person to see this clearly. He was in active touch with Gerhard Hochschild, who was then a student of Chevally at Princeton. Eilenberg suggested that there ought to be a cohomology (and a homology) for algebras. This turned out to be the case, and the complex used to describe the cohomology of groups (i.e., the bar resolution) was adapted to define the Hochschild cohomology of algebras. Eilenberg soon saw other possibilities for homology, and he and Henri Cartan wrote the book *Homological Algebra*, which attracted lively interest among algebraists such as Kaplansky. A leading feature was the general notion of a resolution, say of module M; such a resolution was an exact sequence of free modules F_i

$$M \leftarrow F_0 \leftarrow F_1 \leftarrow F_2 \leftarrow \dots .$$

Earlier work by Hilbert on syzygies suggested this idea; an essential feature was a theorem comparing two resolutions used to prove that the cohomology they give is (up to isomorphism) independent of the choice of the resolution. The bar resolution was not the only one. Any resolution could be used to calculate derived functors.

I took a lively interest in the development of homological algebra. At one point, in a general joint program of the Universities of Frankfurt (am Main) and Chicago, I visited Frankfurt and lectured there on homological algebra. One of my auditors drew a picture of my lecture style—I stand there holding a machine gun, from which a long, exact sequence spurts. Those were the days. The picture still hangs in my office.

The Alexander von Humboldt Foundation provided funds for scholarly visits to Germany. On two occasions, I held such fellowships for semester-long visits to Heidelberg, where I was in active contact with Abrecht Dold and Dieter Puppe. Dorothy and I much enjoyed our stays there; the Foundation even provided a car, a low-powered Mercedes. We visited the castle, toured along the Neckar River, and visited other sights. I lectured on homological algebra and used these lectures to help write my book *Homology*, which compactly summarized the whole subject and included a chapter formulating my definitive understanding of those once-mysterious spectral sequences. I lectured on categories on a later visit to Heidelberg.

I much valued the support of the Humboldt Foundation; at one point, one of my students, John MacDonald, also applied to that foundation for a fellowship. In writing a letter about his qualifications, I chanced to mention that they had once given me a fellowship to study at Göttingen. This news fascinated Heinrich Pfeiffer, the director of the foundation. He wrote to explain that though the name was the same, it had not been the same foundation—the earlier one had gone extinct with the Nazis, but the name lingered on. Such is tradition, perhaps quite independent of the changing political climates.

Chapter Forty Two
Categories: La Jolla & Columbia

After the war, category theory advanced, though in the beginning, progress was slow. But new concepts came into use, notably, the notion of a pair of adjoint functors, formulated in 1958 by Daniel Kan, then at Columbia University. At Columbia, Eilenberg collected a lively and remarkable group of people interested in category theory in 1960, including E. F. Assmus Jr., Michael Barr, Hyman Bass, Peter Freyd, John Gray, and William Mitchell. There were also nine students studying categories there: H. Applegate, Jon Beck, Leonard Charlap, Chester Feldman, William Lawvere, Fred Linton, Saul Lubkin, and Myles Tierney. This group created a concentrated environment for ideas about categories.

In 1965, there was an effective conference on Categorical Algebra held at the splendid seaside campus of the University of California at La Jolla. There were at least 22 speakers whose talks were presented in the published volume, which began with a manifesto by Lawvere called "The Category of Categories as a Foundation for Mathematics." The conference acted to encourage the growth of the subject. It was funded by the Air Force Office of Scientific Reasearch (AFOSR), an office which, thanks to the program director R. J. Porter, had already awarded a number of research grants to experts on category theory. Unfortunately, at the time of the conference, Porter had to announce that his office was unlikely to support further research or additional conferences. This was a change in the general policy of the AFOSR; category theory is very general and quite abstract, which makes for its force in mathematics, but does not fit well into political decisions and priorities.

Among the influential talks at the conference were those given by John Gray on fibered categories, Freyd on stable homotopy, Barr/Beck on acyclic models and triples, and Eilenberg/Kelly on closed categories. These are a sampling of the ideas started in La Jolla. Max Kelly went on to found a lively school of category theory in Australia; the same year, Eilenberg and John Moore published their influential paper "Adjoint Functors and Triples." The basic idea in the paper concerns a pair of adjoint functors $F: X \to A$ and $U: A \to X$; think of X as sets, A as groups, and F_x as the free group generated by the set x, while U is the functor that forgets the group structure on the underlying set. This pair of functors determines an endofunctor $T = UF$ on the underlying category $\mathcal{X}$, while the adjunction produces two natural transformations $\eta: I \to T$ and $\mu: T^2 \to T$ with certain simple properties.

This triple of things (T, η, μ) on a category $\mathcal{X}$ led to the name triple as used by Eilenberg-Moore. It was an unfortunate choice of name, since many other mathematical ideas came in triples, and I later proposed the name monad. Whatever the name, the idea is a powerful one: From the monad (T, η, μ) on the category $\mathcal{X}$, one can reconstruct the upper category A as the category of algebras for (T, η, μ), thus giving a clear formulation to a structure basic to universal algebras where $\mathcal{X}$ is the category of sets and F assigns to each set the corresponding free algebras. Unhappily, universal algebraists have not followed up on these clarifying notions.

Developments in category theory were further stimulated by other international conferences, such as Triples and Categorical Homology Theory, organized in Zürich by Beno Eckmann in 1966, and Category Theory, Homology Theory, and their Applications, organized by Peter Hilton at the Battelle Institute in 1968. These conferences and activities encouraged the growth of research activity in category theory: Eilenberg's original notion that our first paper on the subject would suffice turned out to be a mistake—the subject was much bigger than we thought and spawned more new ideas. At Chicago, my own research was supported by various grants from the NSF and AFOSR, which brought long-term visitors in the subject area such as Jean Benabou from Paris. I had first met him when he

Chapter Forty Two — Categories: La Jolla & Columbia

came to St. Andrews in Scotland while I was a principal lecturer at a summer conference there. While at Chicago, Benabou developed the basic idea of a two-category with objects, arrows, and 2-arrows:

$$\bullet\,,\quad \bullet\longrightarrow\bullet\,,\quad \bullet\Downarrow\bullet\,,$$

satisfying suitable axioms. His work was published by Springer in one of the volumes of my *Reports of the Midwest Category Seminar*, and has led to the now active subject of n-categories, where there may be not three, but n such primitive concepts.

To summarize, category theory started with an abstract description of some basic mathematical structures that arose in the practice of algebraic topology. It then developed that essentially the same type of abstract structures were applicable in other parts of mathematics, and that there were more structures to be found and formulated, such as monads (triples), universals, and higher-dimensional categories. Thus, the subject was continually driven by new discoveries and new analyses, aspects that were not known to us when we first found categories present in homology theory.

Chapter Forty Three
Geometrical Mechanics

In 1967, the distinguished Chicago astronomer, Subramanyan Chandrasekhar, took up the study of relativity theory, and observed at once that its modern form would require a good deal of mathematics, especially differential geometry. At his request, I gave a two-quarter course called geometrical mechanics that was attended by physicists and mathematicians. I gave the same course for several successive years; in the first year, some of my students, primarily Raphael Zahler, prepared extensive mimeographed notes with the assistance of a grant from the Office of Naval Research. Originally, I planned to use a then-new text, *Foundations of Mechanics*, by Ralph Abraham, but it turned out to be too murky and incomplete for my taste. Instead, I had to depend on my undergraduate study of mechanics with E. W. Brown at Yale and my more recent study of differential geometry, in which I was guided by my Chicago colleague S. S. Chern.

My course on mechanics began with the slogan "kinetic energy is a Riemann metric on configuration space," followed at once by examples of a Riemann metric and of a configuration space (the sphere). The course continued with discussions of tangent bundles, Lagrange's equation, Hamilton's principle, and the Legendre transformation, followed by a description of tensors, manifolds, and exterior forms, and then Hamiltonian mechanics (a beautiful subject) and simplectic manifolds. Next came submanifolds of constant energy, vector fields, the Hamilton-Jacobi equations, covariant derivatives, and parallel transport. In one version of the course, I

Saunders lecturing, ca. 1970

entered into relativity theory, where I managed to show why light rays passing the sun were bent by the sun's gravity. I much enjoyed this, and I think my students did too. It was an experience to teach a new course on subjects quite new to me.

Lawvere was an assistant professor at Chicago when I taught a version of this course; from listening to my use of differential geometry, he came to think about the justification of older intuitive methods (S. Lie, et al.) in geometry. This, in turn, led him to start to develop the subject synthetic differential geometry, which uses a version of actual infinitesimals to formulate geometric ideas. His current Ph.D. student, Anders Kock, took up the idea, as did others, including my student Eduardo Dubuc: there was a full-blown development of their ideas, which do indeed provide a working system of infinitesimals for geometry, as presented, for example, in a book by Mordijk and Reyes. However, those promising ideas have not yet entered into everyday use by differential geometers.

Chapter Forty Four
Outdoors on the Indiana Dunes

The early vacations I spent with my mother at Marjorie's cottage on the beach in Westport, Connecticut, had started my interest in swimming and sailing, and Dorothy shared these interests. Later, when Marjorie wanted to give up care of her cottage, she sold it to Dorothy, which made it possible for us to use the cottage freely in the summers, and for a couple of summers we drove east with our daughters to spend a month there. But soon we wanted an escape that was much closer to home, which led us to think of the Indiana Dunes on the south shore of Lake Michigan, about an hour's drive from the South Side of Chicago. We investigated buying a cottage there; the first one we considered was a house right on the beach, between the lake and the lakefront road, in Beverly Shores. But we could not negotiate acceptable terms with the owner, so we gave up. Two years later, the lake level rose considerably, and the house was washed away in a storm. Happily, it did not carry with it our invested money.

Later, we thought of joining with one of my colleagues to purchase a common interest in another house in the Indiana Dunes, but this time, the problem of sharing use of such a house did not work out. In the meantime, I found out about a stretch of beachfront that was completely undeveloped, and the owner rented out space on short-term leases, so vacation seekers built shacks there for summertime use. I found one to rent for the entire summer, a three-room shack with a well, located right on the front dune. That summer (1956)

The Sixties and Beyond

Dorothy at the Dunes, 1964

there was a conference on algebraic topology at Chicago, so I invited the conferees to come out to the dune hideaway. It was half a mile to the nearest road, so one had to walk in on a trail over the dunes carrying all provisions; when we got there, the view was fine and the well provided clear water, but the necessary outdoor facility had collapsed. The famous topologist at Cambridge University, Frank Adams, had come out with me, and I found out that he understood not only topology, but carpentry as well. There was lumber available at a vacant house nearby, and Adams rebuilt the outhouse. Yes, topology is a practical subject.

A year or two later, that whole stretch of pristine dune land was sold to the Bethlehem Steel Company—they flattened the dune, outhouse, cottage and all, and built a huge, modern steel plant next to a manmade Indiana harbor. They also created a new town there called Burns Harbor, and this activity reduced the extent of real dune land available for recreation. Dorothy and I then became interested in the nearby gated community of Dune Acres, which consisted of a hundred odd houses and cottages on an unspoiled stretch of dunes. We talked to the local real-estate agent, who told us about a small house there, right atop the first dune with a wide view of the lake. We rented the house for the winter and found it fine for weekend excursions, and it was also a peaceful

Chapter Forty Four ~ Outdoors on the Indiana Dunes

place to work on mathematics. It was well-built, but owned by a widow who had good reasons to sell it, so we bought the house for a reasonable price. Ever since, we have used it for a weekend retreat—it is a one-hour drive from our Chicago apartment—and as a summer vacation house. I swim in the summer and explore the trails through the dunes in the fall in winter, or, I sit in my large, elegant study and contemplate algebra and categories.

Chapter Forty Five
Categories at Work

My extended collaboration with Eilenberg resulted in 15 major joint papers and four major mathematical ideas: homotopy structure of spaces (Eilenberg-Mac Lane spaces, Postnikov systems), homological algebra (e.g., the cohomology of groups), category theory, and simplicial sets. Of these ideas, category theory has had the widest appeal, and I have had many opportunities to lecture on categories, including the following occasions: the 1964 Colloquium Lectures of the American Mathematical Society; at Heidelberg University with support by the Alexander von Humboldt Foundation; at Kings College London on invitation from Professor A. Frohlich; at St. Andrews in Scotland; at Canberra, Australia; at Bowdoin College during lectures sponsored by Dan Christy and the National Science Foundation; and at Tulane University, thanks to Paul Mostert and the Ford Foundation.

In 1971, I put my ideas about category theory together in a book, *Categories for the Working Mathematician*. As the title suggested, the book aimed to emphasize the use of category theory in various branches of mathematics, not only in algebraic topology. The unusual title did receive a bit of heckling—Frank Adams once spoke of "categories for the idle mathematician." But my emphasis did present the use of categorical ideas for the formulation and clarification of various mathematical structures, not just those of algebraic topology, where it had arisen.

In category theory, the three basic notions of category, functor, and natural transformation appeared in definitive form in our original

1945 paper; Daniel Kan added a fourth categorical idea—adjoint functors—in 1958. Additional categorical structures include abelian categories (Mac Lane), closed categories (Eilenberg-Kelly), the categories of fractions (Zisman), two-dimensional categories (Benabou, now extended by others to weak and strong n-categories), and topos theory (Grothendieck, Lawvere-Tierney). There are also a number of connections to other parts of mathematics, such as algebraic geometry (Grothendieck et al), universal algebra (Lawvere/Manes), logic (Makkai/Reyes and Dana Scott), and operator algebras and computer science (many authors). I can imagine that there are other uses to come, such as developing issues in quantum groups.

By now, there are numerous texts and books on category theory. My own 1971 book aimed to emphasize the manifold uses of categories; it has succeeded in a modest way and is now in a second edition. There are also many other books on the subject, including a three-volume handbook of the subject. Perhaps the most widespread preserve of category theory is in algebraic geometry. The most radical aspect is Lawvere's notion of using axioms for the category of sets as a foundation of mathematics. This attractive and apposite idea has, as of yet, found little reflection in the community of specialists in mathematical logic, who generally tend to assume that everything started and still starts with sets.

The notion of simplicial sets arose from the use of Eilenberg's version of singular homology, which was developed in order to define homology groups for arbitrary spaces that were not polyhedra. For this purpose, think of a polyhedron comprised of simplices (vertices, edges, triangles, tetrahedra, etc.) pasted together. Hence, singular homology started by using singular simplices that were just continuous maps of a geometrical simplex (triangle, tetrahedron, etc.) into the space at issue. This had the advantage of giving homology (and cohomology) groups for an arbitrary space. Solomon Lefschetz presented one version of this theory in one of his books on topology, and Eilenberg soon realized that there could be a more specific formulation of this homology by using only simplices in which the vertices are linearly ordered. Thus, an n-dimensional simplex σ would have $n+1$ vertices, σ_i, $i=0,1,...,n$. The faces could then be numbered,

Chapter Forty Five ～ Categories at Work

Solomon Leftschetz, 1964

where the ith face $F_i\sigma$ is found by omitting vertex i, while the ith degenerate face $D_i\sigma$ has the vertex i doubled. The boundary of the simplex σ can then be presented as an alternating sum

$$\partial\sigma = F_0\sigma - F_1\sigma + F_2\sigma + \ldots + (-1)^n F_n\sigma,$$

which was formulated carefully in Eilenberg's 1944 paper "Singular Homology."

When Sammy and I came to study the space $K(\pi,n)$ with only one nonzero homotopy group π_n, we used this singular homology, with both faces and degeneracies and the relations between their composites ($F_i F_j$, $F_i D_j$, etc.). In the course of our elaborate computation of the homology of such spaces, we spoke of F–D complexes, which were sequences of abelian groups C_n with suitable operators $F_i: C_n \to C_{n-1}$ and $D_j: C_n \to C_{n+1}$. Eilenberg and Zilber later reformulated this useful computational task with different notations (d_i for F_i, s_j for D_j). The explicit definition (see my book *Categories for the Working Mathematician*) gives formulas for composites such as $d_i d_j$. This notion of simplicial sets provides an abstract and combinatorial definition of a space that turned out to be useful in many ways.

Part Eleven

National Science Policy

Chapter Forty Six
As President of the AMS

In 1972, the American Mathematical Society elected me as president for the customary two-year term. By this time, it had gradually become clear that the objectives of the Society in the encouragement of mathematical research meant that the Society ought to be active in the field of science policy in addition to its traditional work in meetings and journals. I proposed expanding the scope of research papers presented at meetings—I recommended to the Council that meetings should include several special sessions that could be initiated by any member on a research topic of his interest. This proposal was adopted, and the result is that each major meeting now has a number of such special sessions that represent the varied interests of research mathematicians and encourage specialists to get together and examine current progress in their field.

As president, I was concerned with science policy. It seemed to me that the presence of two mathematical organizations, the Association (the MAA) and the Society (the AMS), tended to divide the interests of mathematicians. So I asked, "Why not combine the two organizations?" I failed in this proposal—there was too much inertia present in continuing their separate activities in research and in teaching. Failing in combining them, I suggested a common organization active in science policy for mathematics. This, I argued, could be done by the establishment of a Joint Projects Board in Mathematics combining the interests of the Society, the Association, and SIAM (the Society for Industrial and Applied Mathematics).

This new umbrella organization, the JPBM was indeed established, initially under my chairmanship, and it continues to be an effective way for these mathematical organizations to express their common interests in the active matter of government policy for science.

Establishing specialized sessions at meetings and working on science policy issues were my central activities as president of the AMS. Just recently, a member of the current nominating committee for the AMS has complained to me that it is difficult to find suitable candidates for the presidency of the Society. Previously, the position was viewed chiefly as an honor; I was apparently the first president to emphasize that the president could be an active spokesman for mathematics. Some more recent presidents have been much more active than I in matters of national science policy, for example, by protesting the sudden action of one university to cancel its graduate work in mathematics.

Traditionally, the secretary of the Society had been a central figure in the workings of the Society. While I was president, Everett Pitcher was secretary, and he was a firm supporter of my efforts. The central role of the secretary has continued—I recall one meeting of the Society where there were lively discussions of a new bylaw that (for some purpose that I no longer recall) specified that an action could be taken by the Society only with a two-thirds majority vote. The question came up: "Why two-thirds?" The secretary answered: "Because it is the simplest number between one half and one." The questioning thereupon stopped—the AMS has been fortunate to have astute secretaries.

While I was president, the office of the treasurer of the Society was about to become vacant (by tradition, the treasurer is often renominated). I was happy to be able to persuade Frank Peterson of MIT to accept nomination to this office. He has recently (1999) retired from that office after more than 30 years of great service.

During my presidency there was also increased attention to the importance of encouraging women in mathematics. Although this important initiative was sometimes marred, the issue of encouraging women in matters of mathematical research has progressed well, especially with the activity of the AWM (the Association for Women

Chapter Forty Six — As President of the AMS

in Mathematics). In 1930, when I was a graduate student, women were encouraged to get Ph.D. degrees in mathematics, but were not often encouraged to do further research. After writing their theses, they were usually sent off to teach, especially in women's colleges. By now, this has definitely changed, with many impressive mathematical research accomplishments by women.

The AMS also has had to face troubles arising from various political movements and agitation; for example, during the anti-communist activities of Senator McCarthy some mathematicians lost their positions, or were threatened but defended—the AMS worked to defend such individuals. Before my presidency, I was chairman of an AMS committee for such a defense when the SDS (Students for a Democratic Society) was active against the Vietnam War. It so happened that I was invited to visit Tulane University for a special session on category theory at that time. The lively and prospering mathematics department at Tulane had just promoted Edward Dubinsky to be a tenured Associate Professor. However, Dubinsky also happened to be the faculty advisor for the Tulane chapter of the SDS. Dubinsky and these students sat in and disrupted the ceremonial drill of a Tulane military science class, which dismayed the Tulane administration, so they made careful plans to keep SDS away from the coming commencement ceremonies. When the day came, so did the rain; the ceremonies had to be moved indoors, where Dubinsky and the SDS students managed to sneak in and disrupt. As a result, Dubinsky was fired, despite his tenure.

These events occurred just before I came on the scene as a visiting professor. The case seemed to me a violation of tenure and so was of concern to my AMS committee. Hence, I made an appointment to see the dean and explained to him that Dubinsky's dismissal seemed to me inappropriate and that it would badly damage the promising development of mathematics at Tulane. I was not able to change the dean's mind; with his knowledge I made an appointment to see the president, to whom I made the same case. After another negative, I even talked to the members of the board of trustees of Tulane. Again I failed. Dubinsky remained fired, but he happily found a position at a different university. I consider that the AMS should, as in this case, be ready to help mathematicians in such circumstances.

National Science Policy

Traditionally, there had been an informal, though effective, general agreement that candidates for the presidency of the AMS should always be members of the National Academy of Sciences, and the intention of this custom was to emphasize the central role of research for the AMS. This tradition, though still in the background, is not always followed literally. But it has had influence. In 1976, Julia Robinson was elected to the NAS (actually on my nomination, because of her insightful research in logic). She was then soon elected to be president of the AMS.

Chapter Forty Seven
Academy Reports

When Detlev Bronk became President of the National Academy of Sciences in 1950, he continued to energize the National Research Council (NRC). In effect, the NRC became the working arm of the Academy, as well as the National Academy of Engineering and the Institute of Medicine after their establishment. Thus, all three academies were represented on the governing board of the NRC.

In my time, the work of the NRC was subdivided into four assemblies that dealt with disciplines—the Assemblies of Life Sciences, Engineering, Mathematical and Physical Sciences, and Behavioral and Social Sciences—and four commissions that dealt with topics—the Commissions on International Relations, Natural Resources, Human Resources, and Sociotechnical Systems, each headed by a member of one of the Academies.[1] Each of these units was subdivided into a number of parts called Boards, Committees, and Offices, some essentially permanent, some temporary, organized to meet a specific request. Details were handled by a considerable professional staff.

Congress, the president's science advisor, and many government agencies requested Academy advice on a wide variety of topics, some involving much science, and others involving little. In a typical year during the period of 1973–81, the Academy issued some 250 reports. The official requests came first to the relevant board or committee,

---

[1] This structure has since been changed to divisions.

then to the appropriate assembly or commission, and finally to the Governing Board for official approval; some requests for reports were encouraged by the responsible NRC staff member. Requests for reports were rarely turned down, but they were often considerably modified in this elaborate acceptance process. From my own long service on that governing board (14 years at one time or another), I suspect it might well have been much more selective—not all of those apparently pressing problems can be illuminated by science, and not all problems proposed are really pressing.

To prepare each report, the NRC selects a committee of experts, academy members or not. The experts are expected to volunteer their time—which they do, to the general benefit. The committee, supported by staff, meets, discusses, consults the literature (if a related literature really exists), and debates the terms and conclusions of the desired report. The committee does not commission new experiments, but assembles and judges the available information and may recommend further research. Some reports involve large questions of policy, some are highly technical, some represent continuing work by standing NRC committees (for example, by the Committee on the Biological Effects of Ionizing Radiation).

A good indication of the character of NRC reports and of some of the difficulties they may involve may be suggested by the following sample list of selected titles:

1963   *The ASHO Road Test: Report 5*

        One characteristic item from this report on pavement research was that a 50 percent increase in vehicle weight makes, statistically, a fivefold increase in pavement wear. It is not clear to me that this startling result has been adequately reflected, then or now, in the level of vehicle taxes.

1970   *Biology and the Future of Man*, edited by Philip Handler.

        This is one of many extensive studies of the state and prospects of individual branches of science, as then made by COSPUP, the Academy Committee On Science and Public Policy; now this committee includes Engineering, and so is renamed COSEPUP.

Chapter Forty Seven — Academy Reports

1971 *Jamaica Bay and Kennedy Airport*, a multidisciplinary environmental study. This study was made on the 1969 request of the Port of New York Authority, and dealt with "Improving the environment in the face of competing needs."

1974 *Lead in Paint*, investigating the problems of children who ingest lead-based paint.
Potential for the Rehabilitation of Strip-Mined Land
DDT in the Marine Environment, a Review of Present Understanding.
The Effect of Herbicides in South Vietnam
This report was very controversial and the review process contentious.
Methodology of Assessing the Likely Effect of a Projected Fleet of High-Altitude Aircraft on the Earth's Climate and the Biosphere

1975 *Environmental Impact of Stratospheric Flight*
Report of the Committee on Motor Vehicle Emissions
Review Committee on the Safety of Outer Continental Shelf Petroleum Operations
Safety of Saccharin and Sodium Saccharin in the Human Diet
Population and Food: Crucial Issues
In Search of Population Policy: Views from the Developing World

1976 *Report on Project Stormfury*
This report was about the seeding of Hurricanes.

1977 *Decision Making in the EPA*
A series of 8 or 9 reports.
Review of the USA-USSR Inter-Academy Exchanges and Relations

1978   *Ecological Effects of a Sea Level Canal* (in the Isthmus of Panama)
The Risks and Benefits of Recombinant DNA Research
Technology, Trade, and the U.S. Economy
The Safety of Existing Dams
An Integrated Manpower Policy for Primary Care (IOM)

1979   *The Continuing Quest: Large-Scale Ocean Science for the Future*
Technological Opportunities in the U.S. Steel Industry
Climbing the Academic Ladder: Doctoral Women Scientists in Academia
The Effect on Populations of Low Levels of Ionizing Radiation
Energy in Transition, 1985-2010

1980   *Science and Technology, A Five Year Outlook*
Health Planning in the U.S.A.
Lead in the Human Environment
Protection Against Depletion of Stratospheric Ozone by Chlorofluorocarbons
This was, at the time, a highly contentious issue.

This list may serve to indicate some of the hard problems involved, their timeliness, and their remarkable variety. In turn, they may indicate some of the many ways in which science may be useful in practical affairs or, on the other hand, how it may have problematic impacts.

Besides the reports listed here, there have been many others, large and small—some have been influential, some have mattered for a brief time when the subject was pressing, and some have been forgotten. They reach various interested publics, because their distribution varies. Those requested by an agency are of course sent to that agency, and may often be distributed to the public, usually by sale through the Academy Press. In recent years, the Academy has initiated a number of reports on its own, independent of any agency request. The officers

Chapter Forty Seven ~ Academy Reports

and Council have recently taken steps to increase the Academy endowment so as to make possible more independent reports.

To the best of my knowledge, no other national academy of science combines its honorific activities with such a widespread net of advice and reports. It does not happen in the French Académie des Sciences, and the British Royal Society has no reporting activity of similar scope. In Eastern Europe, the various academies have functioned more as congeries of research institutes.

I have not been able to find out why this sort of activity has developed to such an extent in the United States—is it our pragmatic temperament or the competing needs of Administration and Congress for helpful or supportive advice, or was it started by the extensive uses of scientists in the two World Wars, then encouraged by far-seeing presidents of the Academy? At any rate, it represents a way of using scientific fact and understanding for the practical problems of the political scene.

Chapter Forty Eight
George Kistiakowski

George Kistiakowski was born in Kiev, Russia, in 1900. At the end of the World War I he served for a time in one of the White Russian armies. When that army failed, Kistiakowski escaped from Russia and went to Berlin, where he soon received his Ph.D. in chemistry in 1925. He then came to the United States, first at Princeton in 1926 and then at Harvard in 1930. At Los Alamos, during the war, he had a large part in designing the explosive charges required for the atom bomb, and in 1959 he became President Eisenhower's science advisor.

In 1939, Kistiakowski was elected to membership in the National Academy of Science and in 1965 became vice president of the Academy and served two four-year terms in that office. Kistiakowski had noted that many reports of the National Research Council did not involve much input or oversight from members of the Academy, so during his vice presidency, he took the lead in establishing two new Academy committees: the Committee on Science and Public Policy (COSPUP) and the Report Review Committee (RRC).

COSPUP, established with the support of Philip Handler, then Academy president, was a committee composed solely of members of the Academy engaged in preparing high-level policy advice for various government agencies. It now has added members of the Academy of Engineering, so is known as COSEPUP; it also includes members of the Institute of Medicine (IOM), although this is not reflected in the committee's title. I was one of the early members of COSPUP.

The decisive reports of COSPUP were carefully reviewed before they were issued. But Kisty observed that many NRC reports were not so carefully checked. Hence, on his recommendation, President Handler and the Academy Council set up the Report Review Committee (the RRC). When first established, it consisted only of Academy members, with Kistiakowski as chairman.

The intent was that every NRC report be reviewed for accuracy and completeness and then modified on the basis of that review by a group other than its author. For most reports, the responsible NRC commission or assembly appointed a review team, while major and possibly troublesome reports were reviewed by a group appointed by the RRC. In 1971, I was one of the RRC reviewers of such a report; Dr. Kistiakowski and I talked by phone with the chief author about the troubles, and I spoke up vigorously about them—they were appropriately modified.

Dr. Kistiakowski's second term as NAS vice president ended in 1973. The age restriction then for officers (now removed, to end the evil of age discrimination) did not permit his renomination. He evidently wanted to keep the new RRC effective, and apparently felt that I might be able to accomplish this, because he took several steps to see that I was one of the nominees in 1973 for the position of vice president of the Academy—my name was put in both by the regular nominating committee and by a so-called Voluntary Nominating Group, which he encouraged. In the result, there were two nominees for this office; when the election came, Kistiakowski campaigned for me and I was elected. In 1977 I was reelected without opposition to a second four-year term (1977-1981). Upon my election in 1973, President Handler (who must have known of Kistiakowski's actions) asked me to serve as chairman of the Report Review Committee. As a result, my activity as vice president of the Academy (1973-81) was chiefly devoted to managing the Report Review Committee, which was a busy and rewarding activity.

Chapter Forty Nine
Report Review

During my time as chairman of the RRC, the committee members monitored the review process of reports by assemblies and commissions and selected for special review by the RRC those reports most likely to involve subtle or contentious issues. Of about 360 NRC reports in a year, perhaps 40 or 50 were given such an RRC review.

Each RRC review was carried out by a reviewing panel chosen by the RRC members and the RRC staff; these panels were composed principally of members of NAS, NAE, or IOM. In fact, this arrangement was intended to provide a connection between the honorific and the advisory functions of the Academies. The reviewers wrote or telephoned their comments in to the RRC office; sometimes, in complex cases, the reviewing panel met to discuss the problems with the report. The reviewing comments, which were not ascribed to individual reviewers, were then sent to the responsible NRC staff and committee. This committee or its chairman responded, usually in writing, describing the changes made, the points that had not been changed, and the reasons therefore. Often this completed the process, but sometimes more negotiation was needed between the chairman of RRC and of the responsible committee. I recall one case where I went to the president of the IOM to explain why a report of IOM went too much to the extreme. In my eight years of experience, there were only two or three disputes that were referred to the president of the Academy, who bore the ultimate responsibility. For example, I appealed the first NRC report on the effects of halo carbons to the president because I considered that it did not sufficiently emphasize the possibly dangerous loss of the ozone layer which the use of halo carbons might cause.

An effective NRC report to the government should exhibit both scientific accuracy and relevance to policy. For this, the RRC drew up a document "Guidelines for the Review of Reports." Below are guidelines from one version, along with comments on what was intended:

1. *Charge. Is the report responsive to the initial charge?*
   Each report was intended to respond to an explicit government request (the "charge"). It is important that each report state this charge clearly and explicitly describe any deviation from that charge. (It turned out that committees might be tempted to express some strong feelings on public issues not really related to the charge.)

2. *Evidence. Are the recommendations based on evidence and analysis?*
   Reviewers asked that conclusions be consistent with accepted scientific theories and be based upon adequate evidence. Sometimes a report may take refuge in statements as to how "the committee feels," despite the well-recognized fact that committees cannot feel. And sometimes a report may go beyond the conclusions supported by the evidence.

3. *Uncertainties. Are the uncertainties in the data recognized?*
   A report may wish to decide, "How big is the effect in question?" But the effect may depend on many inputs, each of them uncertain, so it may well be that the total uncertainty can be very large—and not so recognized.

4. *Clarity. Is the report clear and concise?*
   A draft report can be nebulous. It may speak of qualities to be measured, without explaining these qualities or saying how they could possibly be reduced to numerical measures. It may

forcibly state plausible conclusions that are really just conjectures. A committee trying to represent all possible views on its subject may write on at great length. (We saw many reports that would have been better at half the length.)

5. *Completeness. Is the report complete?*
   A really complete report may require emphasis on some question not recognized as relevant by the charge or by the responsible agency.

6. *References. How does the report relate to published material on its subject?*
   Reference to other publications is clearly needed, and may often require critical comments on the accuracy or dependability of those other documents.

7. *Fairness. Is the report fair?*
   A government agency may have requested a report or study of its projects chiefly because it hoped for expansion, so just wanted a document arguing for a bigger budget line.

8. *Policy. Are policy questions handled with proper care?*
   NRC reports can and do sometimes provide a firm base for some policy decision. But such policy decisions are best left to the government, even when the committee members "know" what the correct policy should be. The NRC committee should only present the options.

In general, NRC reports should not inject unsupported value judgments into the documents, but should aim to use the best available scientific and technical information to reach clearly supported recommendations to comment on the choice between different contending purposes.

This formulation of eight questions is adopted for reports requested by government agencies. However, many of them may be applicable generally to scholarly work. Thus, the important questions are:

1. Charge—What did you really intend to study?
2. Evidence—Have you really supported the conclusions drawn?
3. Uncertainties—How accurate are those enthusiastic figures?
4. Clarity—Can the reader easily understand what you meant?
5. Completeness—Have you omitted a relevant consideration?
6. References—Did you explain why you do not agree with Dr. X?
7. Fairness—Did you gloss over that counter argument?
8. Policy—If your ideas were applied, would they really work?

As Chairman of the Report Review Committee, my considerable activity in monitoring reviews of NRC reports took a good bit of time and energy, but in the process I learned a good deal about methods of applied science and mathematics. Below I detail some of the methods I encountered.

Cost-Benefit Analysis

This is often used to provide specific data for difficult problems. A striking example appeared in a 1974 NRC study *Air Quality and Automobile Emission Control.*

Clean air cannot be purchased on the market, so the benefits of clean air were measured by "shadow" prices found from property values on the general grounds that homes in a region where the air is clean should command higher prices than comparable homes where the air is polluted. More specifically, the prices of houses in various sub-regions of greater Boston were expressed as a (linear) function of some fourteen different variables thought to influence these prices: clean air, proximity to schools, good transportation, proximity to the Charles River, and so on. The coefficients in this linear expression of house prices were then determined (by regression) to give the "best" fit for the data that had been collected. The coefficient for the contribution of clean air was then said to give the shadow price of clean air.

This particular measure is, in my judgment, exceedingly uncertain. It is not clear that the regression equations used were subjected to any reasonable tests or evaluations, nor is it clear that the list of fourteen variables used in the regression is really an adequate list of variables affecting the price of real estate; if some relevant variables were to be omitted, the coefficient used for other variables (in particular, the shadow price of clean air) could be substantially distorted. It is not clear that the most relevant variable—that representing the cleanliness of the air—is really observed as intended; when buying a house, many people may not be especially aware of the state of the air. Finally, a considerable uncertainty is introduced by aggregating the estimates so found for houses in a local community and extending this to all of the U.S.A.

Regression

Many quantities of interest for determining government policy will vary from time to time, probably in accordance with some dependence upon other quantities. Often the form of this dependence is not known, and one may not even reasonably know all the other quantities that are of influence. Cases like this are often treated by the method of regression: the dependent quantity u of interest is assumed to be a mathematical function of specified (and, hopefully, measurable) independent quantities x, y, z, and so on; one assumes a specific analytic form for this functional dependence. The analytic form is often a linear expression

$$u = a + bx + cy + dz + \ldots .$$

Given such an expression and corresponding data for related sets of all the quantities, mathematical methods (technically, the method of least squares) will yield the theoretically best values of the as yet unknown coefficients $a, b, c, d, \ldots$.

Used cautiously, this approach can be helpful. Frequent use has meant that there are available computer programs to quickly determine the coefficients—with the result that the method is often used when the assumed model is not known to fit well or even when it doesn't fit at all. The essential trouble is that the assumed linear dependence of u upon $x, y,$ and z may be at fault. For example, the intended linear relation may hold only for small changes in $x, y,$ and z, while the real relation over the range of variables needed may be a much more complicated one. The variables $x, y,$ and z may not be independent. Moreover, other important variables may have been hidden or wholly neglected in the formulation of the problem. When this is the case, the formula—and in particular, the values of the coefficients—can be very distorted. This can be quite troublesome, especially in the many cases when the coefficients themselves are used subsequently as the shadow prices in a cost-benefit analysis. (There are striking cases where regression can give dependable values for the variable u, but erratic values for the coefficients.)

Mathematical Models

In recent years it has become fashionable to dignify the application of mathematics to science by describing it as the use of a mathematical model. The term can be a precise one: a mathematical model proposes to replace a given empirical situation by an axiomatic structure, where specified quantities, numbers, or concepts are assumed to satisfy certain axioms (such as equations). Then, the consequences deduced from these axioms can be applied to the given empirical situation once the input quantities can be accurately assessed.

Sometimes such a model can provide a considerable clarification, and at other times it may not at all fit the empirical situation. The RRC has seen many reports where the model is almost certainly too simple for the actual circumstances. The RRC has also seen cases where the use of a battery of mathematical equations served chiefly as a smoke screen to persuade the hasty reader that the authors knew everything. Fortunately, NRC reports have not exhibited the most extreme case of misuse of mathematical models, as in the notorious report on "Limits to Growth" issued by the Club of Rome.

Projections

Discerning what the future will be has forever been a challenge to mankind. What once might have been tried by oracles, divination, or *belles lettres* (as in George Orwell's *1984*) is now more often referred to science. Many scientists have tried to make predictions, only to find that in most cases, what occurred was not what had been predicted. So it is that in recent years the too-ambitious word prediction has been replaced by projection, and so it has come about that the committees have been asked to project all manner of futures, from future manpower needs to future consumption of energy. After due examination of the results, the RRC might conclude that the change of wording from prediction to projection has not resulted in any real advantage in method or result. All too often, what may have been intended as a modest projection is taken to be a prediction.

Manpower problems often suggest projections. A massive report *Manpower for Environmental Pollution Control* was Volume V of the even more massive 1974–1977 *Analytic Studies for the U.S. EPA*. The authors of the report wanted to "assure that adequate numbers of trained and experienced people will be in the right place at the right time," but at the same time had to observe that "it is difficult to foresee with much exactness the flow of qualified workers into the environmental monitoring business," and that "it is risky to predict local government environmental manpower needs." In short, the future is and remains opaque.

In 1978, the IOM issued a report, *A Manpower Policy for Primary Health Care*, dealing with hard questions about the supply of physicians, nurse practitioners, and physician's assistants. One thrust of the report was to find a strategy to increase the supply of physicians in primary-care disciplines—the emphasis on primary care (as opposed to medical specialization). Such cases also illustrate the inevitable temptation to project results that will favor those aspects of the future that the projector prefers. In short: When is a projection a disguised wish list?

The Delphi Technique

All of us have heard of the Oracle at Delphi, which neither conceals nor reveals, but tells in parables. Some years ago, a philosopher of science, once a student of logical positivism, was inspired to invent the Delphi method for problems of policy, which consisted of trying to get estimates from experts of desired facts about the future. The procedure goes, roughly speaking, as follows: send experts a questionnaire as to what the future energy demand will be in the year so-and-so; collect their answers. Codify the answers and send them out to all, so that each expert can see what the *other* experts think. In view of what the other experts think, ask each one to make a second (and arguably better) estimate. The average of these is the final result. There are all sorts of variations on this Delphi method. Indeed, it was tried in one NRC report.

Contemplation of this method reveals that it is not science, but speculation. The merit of the speculation depends altogether on the insight or the care of the speculators, as well as on their subservience to the opinion of their colleagues—or perhaps on their lack of subservience. After inspection of this example, I recommend that the Delphi Method not be used in anything claiming to be science: It fails to meet the proper demand of the Vienna Circle (logical positivism), that science is that which can be verified.

Opinion Surveys

Many questions about the habits, behavior, and opinions of large groups of people are difficult to answer from any regularly available data. For this reason, such questions are often attempted by sample surveys of a portion of the population at interest. In some branches of the social science, survey techniques are almost the only available means of producing the desired empirical data. What can be found out by such surveys depends very much on the circumstances and the care used in building the survey instrument. Survey questions about matters of fact can usually produce fairly dependable answers, though even here questions about the age of the respondent are likely to give somewhat slanted answers (oh, to be 29 again!). On the other hand, questions about opinions are likely to be much more troublesome, because the order in which the questions are asked, or the words with which they are phrased may very well influence the tone of the answer. Moreover, it is not always easy to choose an adequate sample of the population intended, or to get an adequate and representative response from the sample chosen. For this and other reasons, opinion surveys are subject to great difficulties.

Nevertheless, the survey technique has been widely adopted, both because of its unique character and also because of its frequent use for various industrial purposes, market surveys, newspaper columns, and the like. Experts on the subject have drawn up careful instructions about how to arrange questions in the best possible order and how to state them in the most balanced fashion. Unfortunately, these instructions are sometimes not very carefully followed. Survey experts like to compare the results of present surveys to results from the past, and this tends to perpetuate poorly-formulated questions.

Experts Recommend Funds

As already observed, committees of experts on a given subject may be tempted to go beyond the subject in expressing political ideas. They are also likely to recommend more government expenditure for that subject (better telescopes, more drilling on the continental margin, more activities in space, more support of aeronautics ...).

One example is the report *Toward a National Policy for Children and Families*, issued in 1976 by the Assembly of Behavioral and Social Sciences. This report studied ways to enhance child development and "to ensure that every American family has an income sufficient to enable parents to provide the basic necessities for their children." After considering various possible methods of raising low incomes, the report concluded that "it is necessary to examine direct redistribution of incomes as a means of ensuring economic security for families with children" and stated the goal as follows:

> No child should be deprived of access to a family living standard of at least half of the median family income level (after tax) for a substantial portion of his or her childhood and family incomes should not fall below the government-defined poverty level even for shorter periods.

The report did not offer evidence for "half," or evidence for the correctness of government-defined levels of poverty. It did observe (with a reference) that "achievement of this goal would increase the share of financial income going to the poorest 20 percent of families from the current 5.1 percent to approximately 10 percent."

Inevitably, experts on children naturally consider the welfare of children first, and so have reasons to recommend better care for children. In this case, the reviewers did not succeed in persuading the committee that income redistribution went beyond their charge.

All these issues—and others—provided substantial and interesting questions during my eight years of activity in reviewing Academy reports. My task in reviewing was in some ways a return to my undergraduate quest for what I called universal knowledge; as I reviewed reports, I felt the need to learn everything I could about each topic. For example, I even became an expert on American highways. My activity during those years kept me busy with this version of universal knowledge and left little time to do research on categories.

One of my colleagues tells me that it is well known during my time as chairman of the RRC, I was "a terror to anyone who wrote a report." Perhaps I did give people a rough time in the review process—I was critical, as were others on the RRC, and as chairman of the committee, I was the point man who had to negotiate changes in response to written comments. I wasn't always compromising, especially when people didn't try to be responsive or explain better why something was included in a document or written a certain way. Committees frequently overreached their charge or failed to present the evidence for a conclusion or recommendation, and I was skeptical of certain methods, such as those I have discussed. This led to occasional disagreements—I was not known for being diplomatic. But I believe that it is important that scientists speak up to assure that science is not misused. I made it my business not to approve anything that I did not fully understand or that I thought was not scientifically correct. I took this duty very seriously.

I spent a lot of time in Washington during my years on the RRC. We would put in long hours, and 15-minute naps on borrowed sofas kept me going. We worked hard, but knew how to relax after a working dinner with the RRC staff (Bob Green and Greta Schuessler); I would end each evening by reading a few poems—I had fun constructing some doggerel for special occasions, the enjoyment of a naive poet. Here is a sample:

Chapter Forty Nine — Report Review

The Gypsy Reviewer
By George David Oswald Eliakim Robert Lowell
(To a gypsy tune from "Carmen")

When the bosses need some report
We check it over so it's not too short

We want the data to be so firm
That all the bureaucrats will have to squirm

Committees sometimes write at large
But never never bother to state the charge

And if the thought is nebulous
Reviewers then will raise a fuss

While all discussion of benefit cost
Is junk, so might as well get lost

You can't compare a loss of life
With 55 delays to see your wife

Next comes the issue of diet and health
Despite the deficit, it outranks wealth

If your arteries are full of plaques
The dietary literature ain't full of facts

The plaques will grow and the blood won't flow
That diet on eggs may make it so

But biochemistry is subtle and slow
And reports won't say what the authors don't know

As for cancer, there is no answer
Initiator or perhaps enhancer

We must explain to the G.A.O.[1]
That the progress is just damned slow

Bomb blast particles drift afar
To make a nuclear winter

But does it really last that long?
A one-dimensional model's wrong

They want reports now done so fast
That their conclusions cannot last

They meet officials face to face
But leave the prejudice in place

Report review is just our dish
To state the truth our only wish

We PRESS[2] to keep committees straight
So Frank won't give them all the gate.

---

[1] General Accounting Office
[2] Frank Press, now former President of NAS, and former chairman of the NRC.

Chapter Fifty
The National Science Board

Vannevar Bush, at the end of his wartime service directing the Office of Scientific Research and Development (the OSRD), wrote a famous report to the president, *Science, the Endless Frontier*. In this report, Bush described enthusiastically the many ways in which scientific research had contributed—and would continue to contribute—to the general welfare, and urged the establishment of a government agency to support such research. In 1950, after various conflicts and hesitations, congress did establish such an agency—the National Science Foundation (NSF). It was, and is, managed by a director appointed by the president. Its work is overseen by a 25-member National Science Board (NSB), of which the director is also a member.

The members of the board are appointed by the president and approved by congress. The board generally is constituted to represent the various sciences, the various universities, and other scientific interests. At the beginning, Marston Morse was a member of the board for the usual six-year term; since that time there has usually been one mathematician as a member. The AMS has, from time to time, recommended potential members; for example, at one time the AMS recommended E. J. McShane or me as a choice—the shorter Mc was appointed. Later on, my name was again proposed, and I then served as a board member from 1973–1981. It was a fascinating experience. The board met in Washington once a month, each time for a couple of days, and we discussed big grants and major new policies that had been proposed—these were usually proposed by the director, and we usually approved of his ideas.

By this time, the general notion of NSF grants for research was well established. The amount available for grants is essentially determined by the annual government budget for the whole NSF. At the NSF, this is first divided up into fields of science (the Divisions of the NSF) and then subdivided into programs. Each program officer sends copies of the new proposals to experts in the field, who provide peer reviews. The program office assembles the reviews, compares different grant proposals, and makes preliminary funding recommendations that are given for further review to the NSF division concerned. The NSB itself rarely considers smaller individual grants, but sets policy; for example, the usual peer review was recently replaced by merit review. I have been unable to find what the difference is, but I suspect that merit review simply means peer review plus comments by NSF staff.

Usually, at the monthly board meetings, I took time to go to visit the program director in mathematics. The chairman of the board told me that this was inappropriate—I continued to visit.

By this time, a major element of grants in mathematics was the provision of summer salary, usually 2/9ths of the academic year salary of the Principal Investigator (PI). An earlier program director for mathematics, Arthur Grad, had encouraged mathematicians to apply for grants with such summer salary. He hoped to see a "Cadillac in front of every mathematician's house," to make it clear to all that mathematics is important. In reality, I did not see many such Cadillacs, but I did note a number of mathematicians living in more impressive houses.

What sorts of people serve on the National Science Board? How well is the board membership balanced? As a sort of informal test, I have made the following tentative classifications of NSB members at four points in time, as listed below. The classifications are only approximate, and of course do not say how active or effective various individual members may have been. My classifications are "scientists" (usually university professors), scientists who are also deans, university presidents, and so on.

Chapter Fifty ~ The National Science Board

University presidents are normally well-represented on the board, but they are evidently not active as scientists. Usually there are more university presidents than bench scientists, but the bench scientists plus the deans always outnumbered the university presidents.

Make-up of The National Science Board at various periods:

	1950–54	1962–64	1974–76	1991
Scientists	5	8	6	6
Scientist/Deans	6	5	5	3
University Presidents	7	9	5	7
Foundation Presidents	2	0	0	1
Industry Executives	4	2	6	4
Director, NSF	1	1	1	1

There is usually an explicit concern with diversity; for example, two representatives from the same university is a situation to be avoided, and there is similar interest in distribution between states and between fields of science. However, there are exceptions: in 1985, when I learned of a proposed new board appointment, I circulated the appended small poem—my authorship had a slight but traditional New England disguise, which was later left off in a publication in *Science*. The poem was also sent to the White House; in response, I was told that the president needed to appoint people to carry out his policy, and added thanks to me for supporting that policy. I don't know about the support, but here is the poem:

The National Science Board

The National Science Board, with an authorized complement of 25 members, now has eight members from the one state of California. A ninth member from that state has just been nominated to the board. The following comment on the situation comes from up there, beyond the pearly gates:

> Science boards need many talents
> To preserve a proper balance
>   In distributing the monies from the tills:

Scientists who come from benches
Engineers who turn the wrenches
    And some people with administrative skills

Astrophysicists and scholars
Who can total up the dollars
    Biologists and chemistry to boot

Those who know tectonic science
Mathematics, plus some clients
    Of the theories that should be built to suit

Scientists on either ocean
Have a very different notion
    As to what new institutions must be grown

Engineers from the interior
Need no longer feel inferior
    In serving or advising on their own

But an imbalance terrific
Now sweeps in from the Pacific
    With new nominations we can scarce afford

So this ditty's here to warn 'ya
That the state of California
    Has far too many members on the board

    –Ralph Waldo Longfellow

Science 10 Jan., 1986

Next, I put the question: How well does the National Science Board function? What should it be doing and what has it in fact done?

The purpose of the board is sometimes formulated in a favorite political phrase: the board is to set policy for the foundation, so that the director and the staff can execute that policy. In this description, the word policy has a high, attractive gloss, like that of several other popular phrases (for example, decision makers). The platonic idea of a policy strongly suggests that a good policy will clearly determine how to act in every case that arises. But in the support of science,

Chapter Fifty — The National Science Board

things are not that easy: a policy to support only cutting-edge science may suffer when it is not clear to anyone what edge is doing the cutting. I suspect that the word policy diverges from what actually happens: the director and the staff, inevitably busy in the practical problems of running the foundation, propose or even carry out, actions that may amount to new policies, and when reported to the board, the resulting policies are adopted, perhaps in modified form.

A central question of foundation policy is: What types of grants are to be made, and how are grant applications to be handled? This policy has changed over time. The beginning conditions are described explicitly in a book giving the early history of the foundation, under the title *A Patron for Pure Science* (England, 1982), which points out that at the start there were several prominent candidates for the position of director of the foundation. One of these candidates was, by experience and inclination, interested in big-science projects. But before the director was chosen, this candidate accepted a position elsewhere. The first director, nominated by President Truman, was Dr. Alan Waterman, a physicist who had been an associate professor of physics at Yale. He was later active in war research and then served as chief scientific officer of the Office of Naval Research from its inception in 1946. The ONR had taken its first steps under a general federal support of scientific research, and so developed several sound policies for maintaining this support. Dr. Waterman carried these policies over to the fledgling work of the NSF.

That history book of the early NSF observes that Waterman preferred grants for individual scientists to institutional grants; this preference became a policy of the early NSF, and it has definite effects. When individual scientists at universities are the grant recipients, they thereby acquire some influence with university administrators. This power, and its effects, is one of the underlying features of the current research scene. Scientists with big grants get attention: some universities, in advertising openings, may specify that "demonstrated ability to get grants" is a qualification for faculty appointment.

Dr. Waterman's policy on individual grants evidently had the support of the initial science board. Some subsequent directors, for example, Erich Bloch, have modified this policy. Bloch, an engineer, had not had extensive university experience and came to the foundation from IBM. He became director of the foundation shortly after engineering became active there. Under his guidance, the foundation decided to support a number of engineering centers, and this was soon extended to centers in other sciences. This was a real policy change from support of individual research to support of both group and individual research. It might be argued that this change was initially a response to the specific needs of many engineering schools, or, in its later extension, to increases in the prominence of collaborative research elsewhere in science. There is a counter argument that group research tends to be dominated by directions from above, and so less inclined to be imaginative.

From my own six years of service on the board, I can recall exactly one prominent case where the board actually initiated a major policy change concerning the confidentiality of peer reviews. Up to that time (about 1975), peer reviews of proposals were seen only by NSF staff (and sometimes by the board). There were rumors and speculations that some peer reviews were unfair or biased, with the result that some applicants were unfairly treated. (As so often is the case, there was little if any concrete evidence for this; it is my clear impression that the issue was more political than factual). At a summer meeting of the board, it was proposed and voted that henceforth applicants for NSF grants would receive copies of the peer reviews on their proposals, with the names of the reviewers deleted. This change continues in force.

It is almost impossible to measure the merits of such a change. Some applicants, inspecting peer reviews, may learn how better to formulate their proposals, but they are unlikely to learn how to make intrinsically better proposals. Some PIs may be annoyed by reviews that seem to them uncomprehending, and some reviewers may try to disguise their authorship, or refrain from some incisive critique.

In crisis situations, the board can play an important role. I recall one notable event (and I am sure there have been, and will be, others). In the 1970s, social scientists had prepared material called MACOS (Man, A Course of Study) for high-school courses on how people really lived, with material on Eskimo life, including mention of the

Chapter Fifty ~ The National Science Board

practice of sometimes leaving superannuated members to die on an ice floe. Some parents, understandably upset by this treatment of (say) grandmothers, objected vigorously to their congressmen, and soon the foundation was in crisis. The pressure was on the director, in part because there had been some first mistakes in this matter by NSF staff. The director was able to bring the whole problem to the NSB, and with advice and backing from the board, the course material was modified and the crisis subsided. In this emergency—the only major one that I can recall—the board functioned well.

Chapter Fifty One
Science Policy

My service on the National Science Board brought me in active contact with the many problems of science policy, for instance, the period during which set theory was proposed as a mathematics foundation for very young children—even those in kindergarten. This led to the story mentioned earlier, in which Johnny's kindergarten teacher told his parents that Johnny really seemed to understand set theory, but he did have some difficulties writing those curly brackets. With this story in mind, and others, I often objected to the excessive use of sets. This was well before Lawvere had shown that category theory could be used as a foundation for mathematics as an alternate to sets (for an example of this, see Mac Lane/Moerdijk, 1996.) It all reminds me of my earlier attempts to persuade Bourbaki to use categories.

A major policy issue was the developing tendency of the NSF to support big research groups and institutes, which eventually resulted in a proposal to start and fund a new mathematics institute, perhaps modeled after of the School of Mathematics at the Institute for Advanced Study in Fuld Hall at Princeton. NSF staff members such as William Pell, William Rosen, Alvin Thaler, and Bill Pasta were enthusiastic about this proposal—they probably saw it as a way to get additional funds for mathematics. On the other hand, many mathematicians were skeptical of this proposal; they considered that it was likely to reduce the funding for individual research mathematicians. Princeton mathematicians Joseph Kohn and Eli

Stein, among others, chiefly led this opposition. At the time, they were working in a building called the new Fine Hall (previously, the Princeton math department had been housed in the old Fine Hall). William Browder, at Princeton, opposed the idea of institutes, while his brother Felix Browder, at Chicago, apparently liked them.

The controversy over the function of the proposed NSF-supported institute raged on from 1980 to 1982, after my own term as a board member had ended. But I enjoyed the arguments back and forth and described them in the following piece of doggerel (which was then circulated widely). At that time, Rene Thom's catastrophe theory was active.

CUSP Catastrophe for Research Institutes
by George David Oswald Eliakim Longfellow

By the shores of deep-six waters
In the fogs of Watergate
There the budget is assembled
There our projects meet their fate.

In the days of Russian rockets
Science was on every docket
In this heady abstract air
Mathematics got its share.

But we are feeling very sad
For we have lost our Arthur Grad
He said math would never lack
To every prof a Cadillac.

Now the prospects are more humble
With the tax revolt we grumble
And with applications fumble
While the dollar values tumble.

While we do not care a fig
For the science that is big
Still a research institute
With new money sure would suit.

Chapter Fifty One — Science Policy

If new money can be had
Still more projects would be glad
Special years and postdocs too
More instructors, now too few.

William Rosen, Alvin Thaler
Never contemplated failure
While Bill Pell could hardly found
An institute on shaky ground.

Slichter thought he saw the double
Of the physicists' great trouble
So he asked and talked around
What opinion could be found.

Then the National Science Board
Likes new thrusts but can't afford
To give any new resources
To Math Instituting forces.

Singer sang in many voices
Raising questions, posing choices
Many memos from Mac Lane
Some were cogent, some inane.

Representatives of Stein
Saw Fuld thumb her nose at Fine
Stein could leave no stone unturned
'til the institute was spurned.

Enter Joseph J. J. Kohn
He could do it all alone
Making use of old Ma Bell
He gave Pell and Pasta hell.

"All our youngsters need," said Joe
"Is toward Princeton for to go,"
Forms and foliations call
Everyone to new Fine Hall.

Trustees met in Providence
Squandered dollars, tortured sense

Then they thrashed the Institute
Without funds it wouldn't suit.

Nothing makes contention louder
Than a double dose of Browder
One on one side, one on t'other
Brother disagrees with brother.

On the shores of big chief waters
In the town of Washington
See the troubles with the budget
That Math Institute has done.

In the end, the new institute won. The NSF announced a competition for such institutes—and two were established: The Mathematical Science Research Institute (MSRI—often called Misery) in Berkeley, California, and the Institute for Mathematics and Its Applications, in Minneapolis, Minnesota. Both have continued to develop, and I have enjoyed visits at each.

Part Twelve

Travels

Chapter Fifty Two
Visits to China

In 1976, I led a delegation of scientists from the National Academy of Sciences to visit the People's Republic of China.

At this time contacts with China had opened up, and our delegation was briefed on the background by the Committee on the Scholarly Communication with the People's Republic of China (housed at the NAS). We learned of the Great Leap Forward in 1958–59 and the Cultural Revolution of 1966-69. Once in China, we visited universities and academies.

The trip was fascinating: the main street of Peking was full of bicycles and buses running past the Forbidden City (the old imperial palace), complete with a water-filled moat and a big picture of Chairman Mao; opposite, there were pictures of Tiananmen Square where people assembled, and later rioted. On one side of the Square was the anthropological museum with its centuries-old seismograph (water pours out on the side when an earthquake is coming). On the other side of the Square was the ponderous Great Hall of the People.

In Peking, I met the famous Chinese number theorist L. K. Hua (also written there as Hua Luogeng). I had previously known him when he visited the West in 1938, and at that time he had already published notable results in number theory. At the time of our visit, Hua was the Director of the Mathematics Institute at the Peking Academy of Science. We learned that he had not only developed number theory in China, but that he had also introduced operations research, using ideas such as the optimal seeking method (OSM) and the internal path method. The OSM is a method of finding the maximum value of an empirically given function. Hua had adopted some Western ideas, and explained them well to many others in China. At one point he was said to have lectured (by a telephone network) to an audience of about 100,000 people. His method fitted the Chinese doctrine that workers are capable of almost any achievement. We were told that at one point during the Cultural Revolution, Hua's work was defended by Chairman Mao and by Chou En-Lai; for once, a respected mathematician in China.

In Shanghai our chief hosts were at Fudan University, where we met with the members of the mathematics department and their chairman, Professor Buqing Su. He had studied mathematics in Japan in the 1920s and had there learned about projective differential geometry, a topic once popular at Chicago, with Professors E. Wilczynski and Ernest P. Lane. In this subject one applied calculus to curves and surfaces in projective space. The subject had lost popularity in the United States, but Professor Buqing had continued to be active in it. But when he spoke to my delegation, he apologized for his early research in this topic. He said, "I wrote more than 100 papers having nothing to do with practice. I did not do what was useful. What I knew was of no help in practice. What you need to do is to begin with a practical problem, dig deeper and bring theory back to practice. Our country is not advanced enough—this kind of practical work is of some help to our country." At this point the younger members of the Fudan department chimed in, "No, we love our old professor." Some time before then, with the advent of the Cultural Revolution, Buqing Su had gone to the Shanghai shipyard and there had applied his knowledge of differential geometry to ship building by helping to design the contour of a ship's hull.

Our delegation did notice that every visit to every facility began with a presentation of the political line, which was essentially the same line at every facility. Hence, in the last visit (to the Shanghai Normal University), we requested (and received) a much abbreviated presentation. Later, upon request, one of our members gave a talk about the character of graduate education in the United States, emphasizing the great freedom of choice in the United States: choice of students by the graduate school, choice of courses by the student,

Chapter Fifty Two ☞ Visits to China

and choice of research topic by the faculty members. The evidence here of individual mathematical talent was so sharp that the responsible Chinese member subsequently assembled the whole delegation and the local mathematicians, and this time, the delegation received a trite statement of the current political line.

In Shanghai we noticed that whenever our cavalcade of cars reached a busy street corner the traffic light always went green in our favor. We finally discovered why: the lead driver honked ahead to signal our coming. There it was—priority in the midst of equality. But it was a fascinating trip.

At the end or our stay, our delegation was received by one of the vice chairmen of the standing committee of the People's Congress. After a photograph of the whole group of 11 delegation members, one U.S. diplomat, and 20 hosts, we moved into another room with comfortable chairs and cups of tea, at which point the vice chairman and I, through an interpreter, had an hour-long public conversation all about our visit. On this occasion, as elsewhere, our hosts were cordial, solicitous of our welfare, quick to point out the evils of those who follow the capitalist road, and anxious to promote friendship and the socialist reconstruction of China.

On the whole, despite the political emphasis, the visit for our delegation went well. We prepared an official report; subsequently, I wrote an article on pure and applied mathematics in China for a 1980 volume on science in contemporary China. It was edited by Leo H. Orleans for the Committee on the Scholarly Communication with the People's Republic of China and published by the Stanford University Press. In the process, I had to contend with some troublesome editorial work, involving flagrantly inaccurate and needless changes in my manuscript.

I made a second visit to China with a 1981 delegation from the University of Chicago. Our purpose was to consult with the Chinese Academy of Sciences about plans to bring Chinese students to the University here. This time, wives were allowed to join the delegation members, so Dorothy did get to go. Walking had become difficult for her, so we rented a wheelchair to speed her steps. One of the many thrills of the trip for her was seeing the

Great Wall—she was lifted up in her wheelchair and wheeled along the top of the wall. The many Chinese tourists marveled at the chair, although it nearly got stuck going through one of the towers, and she eventually had to proceed on foot when the grade became too steep for any chair.

On this visit, I saw many Chinese mathematicians from the previous visit, with some additional people. One was S. D. Liao, a professor at Peking University who had studied in the early 1950s at the University of Chicago. In that period he had published research papers in topology; in China, he was active in research on dynamical systems. I also met Wu Wen Tsun, who is famous in the West for his contribution to topology, for example, with the so-called Wu classes used to measure obstructions; he was now a member of the Institute of Systems Analysis of the Chinese Academy of Sciences. This shift of his interest appeared to be in keeping with the Chinese emphasis on the practical needs of the people.

Our delegation also visited Fudan University in Shanghai. When there, I asked to see Buqing again. He was too busy to meet with our delegation, but I was able to see him the next day, for an hour-long interview. He was then 79, president of Fudan University, a standing member for a five-year term of the People's Congress (which meets in Peking) and a member of the Chinese Academy of Science (which had about 100 members), as well as a member of the committee that decided on the granting of all higher degrees in mathematics. He said that for the work of these committees, he spent much time traveling to Peking. He had seven vice presidents to assist him in administering the university at Fudan. In addition to all this, he had recently completed a book on computational geometry (with application to ship-lofting). Clearly Buqing had prospered. But he also explained to me that on my previous visit all the things he had told me were wrong; during that visit he had been let out to see our delegation only because we were a delegation from abroad.

Those two exciting visits to China were not my only excursions to Asia. A year or two later I visited Japan on a mission for our academy to interview various authorities and administrators about how Japanese science was managed. I don't think I found this out, but I

Chapter Fifty Two — Visits to China

did write a report. Earlier, I had enjoyed a more mathematical visit to Japan to attend a conference in a splendid setting on Lake Biwa, where I met many accomplished Japanese mathematicians.

Chapter Fifty Three
Anniversary at the Dunes, 1983

When the summer of 1983 came, it was the 50th anniversary of my marriage to Dorothy. Back then, I was an impecunious graduate student trying to finish up my Ph.D. thesis at Göttingen. Dorothy had come over from New York to type my thesis, and we were married on July 21, 1933—a rather hot day—by a German official in the Standesamt of Göttingen. Dorothy recalled that, since the day was warm, I had worn old-style white knickers and she had worn a white summer dress.

Since then, many things had gone well for us. Our two daughters, Gretchen and Cynthia, had grown up; after effective schooling at the Laboratory School of the University of Chicago, they had both graduated from college—Gretchen from Mount Holyoke, where my mother had graduated early in the century, and Cynthia from Cornell, where I had taught for one year. Both daughters married, though Cynthia's husband, William Hay, died shortly after their son Bill was born. Cynthia then moved to London, where she attended the London School of Economics, and later became an editor. Gretchen became a master dancer; for a time she lived in Provincetown on Cape Cod, and later moved back to her favorite city, New York, and married George Vlachos. Dorothy and I often visited both daughters. On our visits to London to see Cynthia, we also enjoyed playing with Bill; at one point I built a cart for him. Dorothy loved to go to the theater in London.

In the meantime, my university career had developed well, with great dependence on Dorothy's help and support. Not only had she

Travels

Saundes with grandson William Vlachos

helped me by typing books and manuscripts beyond number, but she encouraged me and enabled me to spend much time on my work. She enjoyed entertaining and traveling with me on my many mathematical trips—to Europe, Japan, Australia, China, and more.

Dorothy's troubles with encephalitis continued, and it was difficult for her to communicate her ideas to me and to our daughters. The encephalitis had progressed to Parkinson's disease, which made it hard for her to walk. She was no longer able to drive a car, which she had formerly done with enthusiasm and diligence in adventure. Some of her good friends—in particular, Maya May—did some of her shopping for her. But in spite of all this, she continued to be cheerful.

During our summers we stayed chiefly at our house on the first dune at the edge of Lake Michigan. When friends and family visited, they enjoyed the relaxed atmosphere of the dune country and the pleasures of hiking and swimming in the lake. In the wooded areas there were many hiking trails, some of them old wagon roads, some just trails, some very faint. I explored them all and drew up a map illustrating them, which I then shared with our neighbors, and I could always be counted on to fashion a good walking stick from branches on the trails.

Chapter Fifty Three — Anniversary at the Dunes

The Jones sisters: Isabel, Dorothy, and Alice

So it was with enthusiasm that we celebrated our 50th wedding anniversary, and it was not just a one-day celebration. It started a week early, when Cynthia and Bill arrived from London to help us prepare for the day. Gretchen came from New York City for five days with George and their tennis rackets. Both Dorothy's sisters and their husbands came—Isabel and John from Minneapolis, and Alice and Dan from Coral Gables, Florida. And from Washington came Greta Schuessler, a friend and associate of mine during my years of work on the Report Review Committee.

The weather was hot—much hotter than the remembered time in Göttingen. Early comers enjoyed swimming and champagne (though not together). The actual day of the 50th was full of festivities—a dozen people at lunch and 27 cheerful good friends at dinner who brought greetings, flowers, and gifts in happy combination. Thanks to everyone's help, the party went off in fine spirits. Dorothy planned it, I cut back the too-luxuriant foliage about the house, the carpenter repainted bald spots, Cynthia organized the house, Bill helped with the shopping, Gretchen and Isabel helped Dorothy move about, and our team of Gretchen, Cynthia, and Greta worked together splendidly.

What began 50 years before was celebrated with the warmth of friends and family. It was a wonderful, harmonious time.

Chapter Fifty Four
Dorothy's Delights

In September of 1984 we made another successful trip with the wheelchair, this time to Moscow, Leningrad, and Helsinki. The occasion was an international conference on algebra and analysis to celebrate the anniversary of the Steklov Institute, the mathematical institute of the Soviet Academy of Sciences. After our trip, Dorothy seemed less active than usual and was briefly hospitalized with tests for a possible Leningrad virus. She lost some of her normal optimism and did not always stoutly declare that she would be walking well again in two months. Walking did go harder, but she kept at it, even when it took her 80 minutes to go two blocks. In December 1984, she had a brief and unexplained loss of consciousness, followed by another short stay in the hospital. Dr. Mark Siegler, her physician, encouraged her to keep active, so she decided to come with me on a trip to Germany in March, even if it meant using the wheelchair she so detested. Once again, when people asked her how she was, she would always reply, "Just fine," though her private memos reflected troubles. She enjoyed writing letters to Cynthia and talking to Gretchen by phone.

That year, we spent Christmas at our wonderful house on the Indiana Dunes. Dorothy wanted a two-foot Christmas tree, so that it would fit on top of the record player. So on the Saturday before Christmas, I hunted hard to find one, only to come back home with a four-foot tree. It would go on a tabletop, but Dorothy feared that it was too large for our supply of 40-year old decorations. Happily, it

"Die Deutche Gruppe," University of Chicago; top row: Harris, Allworhty, Goodman, Albrecht; middle row: Keck, Kirchner, Hath; bottom row: Puppe, Mac Lane

turned out that we did have enough, and the tree turned out to be especially pretty. Each night after supper we checked the bulbs (they were the old-fashioned ones that darken the whole string when one goes out), then lighted the tree. We watched it all during dinner. Dorothy took wonderful pleasure in that tree, more than on any previous Christmas.

Christmas day went very well; we both enjoyed our presents—books and a necktie for me, and for Dorothy, a new World Almanac and a leather purse small enough for her to easily carry. Dorothy did wish that she had been able to go out to the stores to get something more for me. For dinner, she exercised her culinary magic in fixing chicken and the rest, with plum pudding for desert.

January went cheerfully, despite the discouraging fact that her new neurologist declined as yet to give her the hoped-for, new, experimental drug for her Parkinsonian syndrome, and instead prescribed increasing, massive doses of Bromo-cryptin. As usual, she was very exact and alert in taking all of her too-many drugs. Every morning Dorothy fixed breakfast, and I fetched the needed drugs; every

Chapter Fifty Four ~ Dorothy's Delights

evening she fixed a succulent dinner. Each Monday morning one of her friends took her to the meeting of the French group. She much enjoyed the dinner at the Quadrangle Club of the faculty wives group, as well as the dinner club called the "X-Nominates" held there. On an earlier club occasion, Steve Stigler gave an interesting talk on an aspect of statistics, a favorite subject for Dorothy.

Thursday night, the 31st, we went to the Court Theater on the campus to see a Molière play. Dorothy had observed that though I had taken her to the Lyric Opera in the fall, I hadn't taken her to the theater. So we went. Dorothy walked from the car to the theater through the snow in her boots, and we both very much enjoyed a subtle play, staying afterwards to hear a discussion with the director, Nicolas Rudall.

Friday, it was breakfast as usual; I got *The New York Times* while she fixed an omelet widely known as the Dorothy omelet, which meant burnt on the bottom and raw on top because she couldn't turn it over. Eating a Dorothy omelet once inspired a vow by Jean Callahan to finish her thesis. The day was cold, so there was some doubt as to whether Maya May's car would start. It did, so Maya, as on every Friday, brought Dorothy her needed groceries from the Co-op. We left for the Dunes a little late, at 4:20. There we had a very pleasant dinner with fish that Maya had obtained for us that morning, as well as a fire in our fireplace.

Saturday morning we rose late at 9. After breakfast there, without *The Times*, Dorothy started making cookies for the French group that she would entertain next Monday in Chicago at our apartment. She had brought down a special recipe book with instructions for making a particular cookie without sugar, to suit one of the members of the group. The day was cold and clear, so I took a brief walk, while Dorothy went downstairs to do some laundry in the machine we kept there. I later brought it up for her. Dinner went well, with another kind of fish and some of her cookies for desert. We had a splendid roaring fire in the fireplace, much better than any before this winter. After Dorothy did the dishes, I read to her a bit from McNeill's *Pursuit of Power*, which she much enjoyed. Then she announced that she would take a bath—the tub there was quite easy for her to get into. I helped her in, drew the water, helped her out, and helped dry her, and so to bed, with moonlight on the snow outside.

Sunday morning, the 3rd of February, Dorothy's walking was worse than usual. She declared, nevertheless, that she had much to do; in particular, she had to write a letter to Cynthia. First she set out to make more cookies, this time with sugar, for the French group. I went out to skate, but found the rink not plowed, so walked along the roads instead. When I came back, I knew that Dorothy would hear me open the garage door and know that I was home, so I stayed in my study downstairs to write a letter to my brother David. When I shortly came upstairs to find some information, I called out to Dorothy. There was no answer; I found her collapsed on the kitchen floor. The ambulance came rapidly. Their valiant efforts to revive her were in vain. Halfway through, they reminded me that a small lot of cookies was burning in the oven.

I remember with gratitude the stalwart help she gave me over more than 50 years, and the beauty and grace that she brought to my life.

Part Thirteen

Advising

Chapter Fifty Five
Chicago Graduate Students

One of the great privileges as a professor is the opportunity to observe and help guide the intellectual development of graduate students. The lively and intense mathematical atmosphere at Chicago attracted many eager and able students hoping to work on a thesis. Directing their theses was an exciting and varied task, as will appear, and it is a pleasure to recall some of them and their work.

In the 1950s, Robert Solovay was an undergraduate student at Harvard, but became restless there and transferred to Chicago instead. I first met him when he was a student in a basic topology course I was teaching. From that start, he decided to work in differential topology. He was a student with drive and talent, and his studies reached far beyond my own knowledge of the field, but I served as his official thesis advisor. After he finished his thesis, he won an NSF postdoctoral fellowship. He had developed a catholic interest in many aspects of mathematics and theoretical physics, and joined the faculty at Berkeley, where he was deeply, and properly, impressed with Paul Cohen's spectacular independence results in the foundations of set theory. Solovay and Dana Scott then jointly devised the method of Boolean-valued models to construct models of set theory, which presented an alternative to Cohen's original construction by "forcing." Solovay provided some spectacular examples of these methods. In my own view, they are also an alternative to the use of topos theory to categorically construct models of set theory out of a special type of category, an elementary topos. Solovay's construction made him famous, and it was a great pleasure for me to observe his accomplishments.

Advising

During the Stone Age, there were no specialists in mathematical logic on the Chicago faculty except for two years when Dana Scott came to Chicago as an instructor. Despite this absence of logicians, there were inevitably some students who wanted to work in logic; I became their guide, perhaps because of my earlier work the subject. The first example is Anil Nerode, who, in 1954, asked me to direct his thesis in logic. I encouraged him, but observed that I was about to take a leave of absence in Paris the following year, which did not at all dissuade him. While I was in Paris, he sent me a draft of his thesis, which was, in effect, a study of an aspect of "closed" categories, or categories where the functor $hom(A, B)$ has the expected adjoint. I was pleased with the results and actually presented them at a logic seminar in Paris. It has also been a pleasure to follow Nerode's subsequent work; he became an expert on recursive functions, and as a professor at Cornell, gave effective guidance to many new Ph.D. students, both in logic and computer science.

Shortly after Nerode finished, Michael Morley, who had been a graduate student at Chicago for quite some time, came to me one day with a proposed Ph.D. thesis in model theory on a topic making extensive use of the well-known compactness theorem. I looked over the thesis carefully and regrettably concluded that it didn't display enough originality. Hence, I told him, "Today such applications of the compactness theorem are a dime a dozen. Go do something better." It was quite an abrupt response, but Morley took a trip to Berkeley, then and now a hot bed of logic and model theory. When he came back, he presented me with a wholly different thesis that used ideas from algebra (transcendence degrees) to construct certain ranks for models. Once I read this thesis, I immediately accepted it, and it established his reputation in logic—those ranks are now the much used Morley Ranks—and got him a position at Cornell. In brief, his was a case in which I dared to make an initial negative decision that turned out to be right.

Next, William Howard, another long-time Chicago graduate student, and also a protégé of Weil, came to me with his intended thesis in logic. I knew that Weil had once suggested a thesis topic to Howard; Howard thought he had solved the intended problem, only to find that an essential lemma was wrong. I believe that Weil had been willing to accept his work despite this trouble, but this didn't seem appropriate to Howard. For a period, then, he worked under Stone's general guidance on a topic in partial differential equations, but no definitive thesis was forthcoming. A general departmental fairy tale went as follows: Student to Stone, "I'd like to write a thesis about linear operators on Hilbert space." Stone to Student, "Why don't you?" At any rate, Stone's tutelage did not lead to a thesis for Howard, but Howard's own natural inclination led him to the draft thesis that he then presented—with no advanced warning—to me. I read it, to find the topic in logic interesting, though a bit incomplete. When Howard came to see me, I told him this, and gave some specific suggestions as to what he might add to the draft. His response was, "No Professor Mac Lane, *this* is my thesis." Knowing something of Howard's previous thesis trials, and knowing something of his ability, I then and there accepted the thesis as is, and the department accepted it. Things turned out well: Howard continued his research in logic and spent a period at Penn State University with the well-known logician Haskel Curry, who was a good friend of mine from Göttingen. At Penn State, they developed a splendid idea of the Curry-Howard isomorphism, or "proof as types."

In this case, I can claim that the decision to accept Howard's thesis was the correct one, but I know of no automatic rule to decide such issues. It seems the best we can do is guess as well as possible, and then worry, wait, and see what develops. Howard is now at the University of Illinois at Chicago, where he has effectively directed graduate research in logic and the history of logic. I value his continued friendship.

At that time, other students of mine were not explicitly working in logic, but in related areas. For example, Arthur Kruse (Ph.D. 1948) studied certain "block assemblages" which really belong to logic. In 1966, David Shafer wrote a thesis in logic, using some of the then new techniques of "forcing" due to Paul Cohen. He understood them better than I did at the time.

Algebra, rather than logic, has been a dominant part of my own research, and this is reflected in the work of several of my students:

Advising

Benjamin Moyls on valuation theory; William Ballard on the cohomology of fields; Edward Halpern on Hopf algebras; Ronald Nunke on the functor Ext for modules; and then, in 1970, David Eisenbud on Torsion modules over Dedekind prime rings. Eisenbud has gone on to do additional important research in algebra and in algebraic geometry. I was much pleased when he recently became the director of the Mathematical Sciences Research Institute in Berkeley.

As I described previously, I had one outstanding student, John Thompson, whose work in homological algebra provided a striking example of what can happen with the guidance of graduate students: It is possible for a student to go far beyond what the professor of the subject could have done. In his case, I was right in guessing that there were real prospects in group theory, and I am delighted that they went so much further than I could have foreseen. As is well known, Thompson continued to publish spectacular results.

There were a number of students who worked in algebraic topology. In 1963, Thomas Hungerford wrote about Bockstein spectra; he has since done further research in algebra and acted effectively as a departmental chairman. In 1960, Robert Szczarba studied the homology of twisted Cartesian products, a natural sequel to the Eilenberg-Zilber theorem. Szczarba has continued his research and for a period was the deputy provost at Yale. In 1964, T. C. Kuo investigated universal objects for spectral sequences, thus reinforcing my interest in such topics. Earlier, in 1960, Arunas Liulevicius factorized cyclic reduced products by secondary cohomology operations and subsequently became a member of the Chicago department of mathematics. From Aarhus in Denmark, Leif Kristensen came to Chicago for graduate work—his 1961 thesis dealt with the cohomology of two stage Postnikov systems, a natural development of the work on Eilenberg-Mac Lane spaces, which are actually one stage such systems. Kristensen returned to Aarhus, where I have also had the pleasure of extended mathematical visits. Also in 1960 Joseph Z. T. Yao wrote on the "Moore-Cartan theorem and the Leray-Serre theorem"—1960 was an active period for my interests in algebraic topology.

Chapter Fifty Five — Chicago Graduate Students

A half dozen of my students worked in category theory. Eduardo Dubuc had come to Chicago from Argentina, and here took an active interest in categories. His thesis (1969) made effective use of Kan extensions in the treatment of closed categories. He has returned to Argentina, where he has made substantial contributions to the use of categories in founding synthetic differential geometry—I value his deep results. In 1969, Paul Palmquist wrote on double categories of adjoint squares; he has subsequently been active in industrial consulting in California.

My favorite theorem asserting that "all" diagrams commute did attract student attention. Harry Dole wrote on such matters and completed a thesis, but did not push to get it improved or published. The decisive contribution came from Rosie Voreadou. Rosie first appeared when Dorothy, Gretchen, and I made a mathematical visit to Greece. My friend there, Professor Zervos, arranged to get us a guide to some of the Greek islands. Early one morning, Gretchen came to our room to report that, "Clytemnestra has come to guide us!" She referred to Rosie, who was then mourning a death in her family. Mourning aside, she took us on a splendid tour of the islands, such beautiful spots then and now. She was fascinated with mathematics, and soon (with a suitable fellowship) came to study in Chicago, where she wrote a wonderful thesis on commuting diagrams; the essential result was a careful enumeration of the diagrams that commute. This result then led to fuller refinements by others, such as the Russian expert, Soloviev. Rosie herself returned to Greece, but has not been able to continue active research there.

At one point my good friend, the topologist Frank Adams, had invited me to visit Cambridge University in England. While there, I gave a series of lectures on the new developments on topos theory started by Lawvere and Tierney, but not yet reduced to book form. Peter Johnstone, one of Frank Adam's students, became fascinated with the new subject, wrote his Cambridge thesis on it and then a text, *Topos Theory*. While he was preparing this influential text he came to Chicago as a visitor. My then student Kathleen Edwards was fascinated and wrote her thesis on a question on this topic; thus, she was both my student and my grandstudent!

Advising

Saunders with Steve Awodey, one of his last graduate students, 2000

The presentation of the fascinating subject of Topos theory was the objective of my next book with Ieke Moerdijk (1993). When I subsequently taught a graduate course on this subject, one of my auditors was a graduate student of philosophy, Steve Awodey. He recognized deep connections between Topos theory and type theory in logic, and his thesis (officially submitted to the Chicago philosophy department) made decisive advances on this topic.

I have even had one thesis student in computer science. When an earlier Chicago committee on Information Sciences was disbanded, Suzanne Ginali was left as an orphan Ph.D. candidate and I cheerfully managed to finish the direction of her thesis.

The history of mathematics offers considerable and often neglected prospects. I have on occasion done some study of the history of algebra and of the influence there of Emmy Noether, and I have helped direct two Ph.D. theses on the history of mathematics. The first developed from the Chicago committee on The Analysis of Ideas and the Study of Methods, a committee managed by the philosopher Richard McKeon. This is the committee that sponsored the topic

Chapter Fifty Five — Chicago Graduate Students

and course "Observations, Methods and Principals"—OMP for short. Once, when McKeon asked me to lecture for his committee, I responded with a lecture "Only Mathematics is Precise" (OMP). Despite my choice, McKeon asked me to be a member of his committee. The result was an interesting Ph.D. thesis for the committee by Joel Fingerman on the history of algebraic topology.

Several years later, Ronald Calinger—a graduate student in history—decided to write on the history of Leonard Euler. His thesis was jointly directed with Alan Debus, a professor of history. Calinger has gone on to be active in developing the scholarly history of mathematics and he has recently published such a text.

The study of Eilenberg-Mac Lane spaces $K(\pi, n)$ with only one non-zero homology group has continued to fascinate me. The famous Henri Cartan seminar on this topic had made decisive advances by introducing a general notion of a "construction" to pass from $K(\pi, n)$ to $K(\pi, n+1)$. Two recent students of mine have added to this work: Ross Hamsher, in 1973, presented an elegant conceptual description of $H(\pi,1)$. Subsequently, Hamsher became a practicing actuary. About the same time, Gerald John Decker made further effective studies of $H(\pi,n)$ in his thesis. Decker later took up classified scientific research. I still consider that much remains to be done in the subject of analyzing $H(\pi, n)$.

Geraldine Brady, originally a mathematics major as an undergraduate, had started graduate work in philosophy with Professor Richard McKeon in the Committee on OMP. As with many of McKeon's students, a finished thesis did not result. Brady talked with my former student William Howard (at the University of Illinois at Chicago), who suggested historical work on the 19th century American logician Charles Sanders Peirce. Brady's resulting paper, with my support, was accepted as a Master's thesis by the Chicago philosophy department. When Brady wished to continue her historical studies on Peirce, I pleaded ignorance of the history of logic and put her in touch with my former student Anil Nerode at Cornell; he had a better grasp of the history of logic and considerable experience in directing theses. Brady's resulting thesis was accepted for the Ph.D. at the University of Oslo, where the tradition in logic started by Skolem is still active.

Advising

There have also been a couple of non-Chicago theses. Once, I was invited to visit and lecture at the Louisiana State University, which was an attractive offer because one of the professors there was H. L. Smith, who was famous for a Chicago doctrine of Moore-Smith limits (Moore is E. H. Moore, first Head of the Chicago department). But when I got to Baton Rouge, I found that Professor Smith had just died, leaving a graduate student, M. C. Wicht, with an incomplete thesis. Fortunately, I was able to oversee the completion of that thesis; Dr. Wicht went on to be a faculty member at a university in Georgia. He was a sailing enthusiast, but to my regret I never was able to accept his invitation to go sailing with him in the Caribbean.

At Chicago, the College math staff had appointed several graduate students to the faculty; sometimes they came with an unfinished thesis. One of them was Howard Stauffer, who had started a thesis at Berkeley on Grothendieck-style algebraic geometry. He found some of the necessary background ideas mysterious and abstruse. That they really were, but I managed to help him to finish his Berkeley thesis on "Completions of Categories, Satellites and Derived Functors." Stauffer is now on the mathematics faculty of one of the California state universities.

Such is my experience with the guidance of graduate students. It is a splendid activity, helping beginning mathematicians to learn exciting subjects and to develop new insights.

Chapter Fifty Six
Friends and Mentors

From time to time, the talk and speculation about science policies picks up new slogans; for example, every problem now becomes a challenge. In 1998, there was considerable popularity in discussions of the mentoring of graduate students. This can be an attractive prospect, with many different aspects, as I know firsthand. But I was somewhat surprised to find it the subject of a recent report from COSEPUP (the Committee on Science, Engineering, and Public Policy, of the National Academy of Sciences). The report was entitled *Advisor, Teacher, Role Model, Friend...on Being a Mentor to Students in Science and Engineering* (ATRMF). In connection with this report, the Academy Home Page carried a Faculty Evaluation Form: Mentoring with the intent that graduate students of science and engineering at universities be given copies of this form (or a modification), to be used by them to rate their faculty advisors, and in particular, their thesis advisors.

This form provides 20 items on which the student can evaluate his or her professor, including, for example, seven points on career development and two on research. Not one of these 20 different points mentions helping a student to get new ideas or build new apparatus, and helping students write clearly is left to the end of the list. In other words, the current practice of student evaluations of teachers in a class is to be extended to the performance of thesis advisors, with the difference that the graduate students are given, in advance, a 20-item checklist of possible professorial negligence; to be more succinct, the complaints about mentoring are provided before the act. This is, to put it mildly, startling. It does not ask the student's opinion, but instead suggests what that opinion should be. The ATRMF report urged a thesis advisor to be a friend. It is not quite clear how a professor can be a friend to a student who has a negative rating sheet in hand, ready to send to the department chair.

The preparation of a thesis with an original contribution to knowledge has been the central aspect of graduate work in universities, in the sciences and, recently, in engineering. The effectiveness of such thesis experience is evident in the current success of American scientific research, and such thesis work is carried out in many different research universities. There have been complaints that some students have not been well advised by their mentors, but the ATRMF report does not give any systematic documentation on this issue. Moreover, although many new Ph.D.s have great difficulty in finding suitable employment, the report gives no evidence whatever that this happens because of inadequate mentors.

I find the contrast between this mentoring query and the active process of graduate study startling. My own experience with graduate students is various—different students have different needs. There is no one formula that tells us how to treat a student. The situation of students in mathematics is by no means typical of that in other fields of science, where the discipline provided in laboratory work and supervision is quite dominant. But the strange idea that one can get better mentors by subjecting faculty mentors to a pre-prepared form is simply ridiculous. It is one of the wholly mistaken side activities of the Academy.

The proper guidance of graduate student theses is a many-splendored activity not to be subjected to bureaucratic rules or guidance from Washington.

Chapter Fifty Seven
Rating Research

"Where does our department stand in the latest ratings?" is now a frequently asked question—the public analysis of the purported relative standing of departments has become increasingly popular. This practice began in 1961 when Carter issued for the American Council on Education an Assessment of Quality in Graduate Education. There were several subsequent such reports. I was skeptical of the accuracy of such ratings and I hoped that some new and really objective rating would be developed, for example, by using lists of which faculty members had been asked to give invited addresses (for example, at AMS meetings). With this in mind, I found myself a member of the NRC committee for preparing a new set of doctoral ratings (those subsequently published in 1982). My hope for objective ratings did not work out: the 1982 report used the accustomed reputational index based on a consensus of faculty opinion, even though the report observed that, "This opinion may well be founded on out-of-date information or be influenced by a variety of factors unrelated to the quality of the program."

Indeed, each NRC rater assesses the "scholarly quality of program faculty" under one of several labels ranging from "distinguished" to "not sufficient"; these words are turned into numbers (!) and then averaged, and this average is the representational rating of the program. The result is a linear order of all the departments in each field. I have come to the conclusion that such linear orders of departments are simply meaningless, and I said so in my article,

"NRC Ratings of Research Doctoral Programs" (*Notices of the AMS, Vol. 43, 1996*). Despite my attempted eloquent argument, everybody continues to rate departments in linear order, as if mathematicians had never developed the idea of partial order. So much for abstract mathematical ideas.

Chapter Fifty Eight
The NAS Research Roundtable

The end of my term of office as Vice President of the National Academy of Sciences did not end my interest in Academy affairs.

Many questions of science policy arose there, and to handle some of those questions, the Academy established a new committee, the Government University Industry Research Roundtable, or GUIRR. The idea was, in Washington parlance, to combine the views of all the "stakeholders" in scientific research; it was also designed to be a forum in which Academy representatives could conveniently meet with government officials.

From time to time, I have examined carefully, and been critical of, reports issued by the GUIRR. For example, a discussion paper from October 1989 entitled "Science and Technology in the Academic Enterprise: Status, Trends and Issues" contained many strange statements that troubled me. Page 1–24 announced, "New disciplines are emerging... This development, however, has not been without problems—most notably the faculty reward system that favors single disciplinary research in established fields." This attack on scientific discipline did not present any discussion of the problems of setting standards with new fields.

Later on the same page, the report asked, "How can new appropriate outcome measures of academic research be developed to evaluate research productivity and efficiency?" Here and elsewhere, the report did not mention the quality of research results as possible measures of achievement. It is not clear to me that productivity is the appropriate objective of scientific research; rather, it is an administrative device.

Evidently, the direction of this report favored administrative ideas. I classified the 15 people who prepared the report as follows:

Government Officials	2
Corporate Executives	3
University Administrators	7
Scientists	3

A 1992 GUIRR discussion paper "Fateful Choices" states (p. 22), "While any changes in the tenure system will be difficult, one such option would be to appoint research faculty to annually renewable, multi-year tenure...Such 'rolling tenure contracts' might give institutions the flexibility to reform their research programs." These criticisms of tenure were not combined with any examination of the importance of tenure in encouraging independence in research work. In other words, the attack on tenure was not based on any careful analysis of the tenure system or any consideration of the bureaucratic effect of annual examination of faculty activities.

A 1994 GUIRR report concerned "Stresses in Research and Education at Colleges and Universities." The report wished to "emphasize research of an economic relevance to the nation and to collaboration with industry." There was criticism of the famous 1945 Vannevar Bush report (*Science, the Endless Frontier*), and an announcement of the need to reconsider the fundamental assumptions of the national science policy (without much analysis of the assumptions).

Other reports from GUIRR display corresponding emphasis on administrative and industrial views of science. It is my general conclusion that it is important that working scientists pay attention to these troublesome trends. I have variously tried to do this. At one point I wrote an article "Should Universities Imitate Industry?" that was published in *American Scientist*, Vol. 84, 1996. Another article of mine, "The Travails of the University," was published in *Perspectives in Biology and Medicine*, Vol. 41, Autumn 1997.

The role of the Academy in these dubious activities is to be much regretted. The Academy should not try to instruct scientists how to act—individual scientists must be continually on the watch to guard against such bureaucratic trends.

Part Fourteen

Later Developments

Chapter Fifty Nine
The Philosophy of Mathematics

My early interest in mathematical logic quite naturally led me to concerns with the philosophy of mathematics. However, I found most of the books on the subject to be misdirected; there was much concern about the nature of mathematical objects—what they are and where they exist, if anywhere. But many philosophical studies by Wittgenstein and others paid little attention to the actual substance of mathematics beyond the most elementary concerns. It seemed to me that one ought to be able to describe more exactly what really is there in mathematics and, on the basis of knowing what is there, which philosophers like Wittgenstein didn't know, to build up what a philosophy of mathematics really ought to be.

With this in mind, I tried to capture in print a description of the form and the function of mathematical ideas. My efforts became articulate in 1983 when I was invited to visit the new NSF-supported Institute for Mathematics & Its Applications at the University of Minnesota; there, I gave a series of lectures about the many topics in mathematics.

The resulting organization of my thoughts on the subject was presented in my 1986 book *Mathematics, Form, and Function*, which aimed to examine some of the basic parts of undergraduate and first-year graduate mathematics to see where they come from, how they fit together, what the relations are between them, and what this says about the possible philosophies of mathematics. I discussed geometry, transformations and groups, calculus, and linear algebra. There was an extensive chapter on mechanics, followed by a chapter describing the wonderful structure of complex analysis, plus a chapter on categories, ending with a description of what I called the mathematical network. I described this network in terms of diagrams (graphs) connecting different topics. I ended with the claim that mathematics results are correct but not true.

I have written other papers on this subject, in particular, one claiming that mathematics is protean in the sense that one and the same mathematical situation has many different empirical realizations; for example, the idea of multiplication (a product) has many different uses. My discussion ended with the recounting of the development of the ideas of Galois theory.

This book and subsequent work was, in part, an attempt to revive interest in the philosophy of mathematics, which is, admittedly, an offbeat subject not highly regarded by most mathematicians. There was a lively interest in the topic in the '20s and '30s, when there were certain contradictions and controversies, and it's been a subject that I've been interested in ever since then, from the time I started studying Whitehead and Russell as an undergraduate. But it was a subject in which it was hard to get ahead, so I took up other interests. It certainly is the case that most mathematics departments wouldn't hire somebody specializing in the philosophy of mathematics—mathematics departments regard the proving of mathematical theorems or getting new mathematical insights as the crucial thing. But I decided I had reached the point in my career that I could think about it. My colleagues didn't mind. The lectures I gave in Minnesota on the topic of the book were well attended by all sorts of colleagues who were interested in my views about mathematics. I also gave a seminar on the philosophy of mathematics, one year at Chicago, in which I put out various contentious issues and got back sass from my colleagues. I'm sure they didn't regard it as serious mathematics, but some did take an interest in it.

For me, it was a pleasure to return to a topic in which I've always had an interest.

Chapter Sixty
Second Marriage

Osa Skotting was born in Denmark on February 18, 1924. She grew up in an active family of three sisters—Eva, Lasse, and Doetter, and one brother—Poul. After the war, Osa came to Chicago. At the time, Emmy Stone (Marshall Stone's wife) needed support and assistance, so Osa stayed for a while to help her with their children, and also came to help Dorothy and our children. She has later complained that I then paid little attention to her—I gave her simply an absentminded nod when she turned up—although I now know her help was important.

Subsequently, Osa worked in Chicago as a commercial artist (she had a Danish degree in design). She married Irving Segal, then a member of the Chicago mathematics department. Irving and Osa soon left to go to MIT. In Cambridge and in Lexington, Osa and Irving raised their three children William, Andrew, and Karen. However, later the marriage came apart and ended with a difficult divorce.

After the divorce, and after resuming residence back in Denmark with Karen, Osa returned to Cambridge. She acquired a master's degree in art therapy, an effective use of her artistic knowledge and talent. Some time later, I was a member of the Council of the American Academy of Arts and Sciences for three years. This entailed trips to Cambridge for meetings. On one such trip, I had the happy idea of inquiring about Osa—and inviting her out to dinner. She later came to Chicago to work for a children's bureau in art therapy. Osa and I had several dates, which led to a great idea: I proposed to her. Happily, she accepted, and we were married on August 16, 1986.

Later Developments

Osa and Saunders on their wedding day, August 16, 1986

Our daughters came for the ceremony, as did Osa's sons, William and Andy. Karen was at that time working in London.

It has turned out very well for us. My second marriage presented a situation wholly different from the first marriage. Back then, I was beginning to make my way in the academic world and Dorothy helped a lot. By the time I married Osa, I had made my way, such as it was. Osa has been a wonderful support—watching my work, listening to my academic problems, making useful comments, exercising together with me every morning (almost), overseeing the taking of my various medicines and, above all, encouraging my research.

Because of a number of mathematical invitations from various parts of the world we have very much enjoyed rewarding trips together. We have met with much warm hospitality wherever we have gone. It was during the first year of our marriage that the National Academy Scholar Exchange Program enabled us to make a trip to the then Soviet Union. First we went to Moscow, where we talked with many people, in particular with the group theoretician Kurosh, whom I had previously met when he came to the United States. I lectured at the Steklov Institute in Moscow and colleagues

Chapter Sixty ～ Second Marriage

Saunders at Tbilisi, Georgia, 1987

arranged for Osa to meet with Russian authorities on child and adolescent care. She was even given a tour of an institution for artistically gifted children.

We were treated to a Bolshoi ballet performance of *Giselle* at the enormous People's Palace, where we sat in the audience with thousands (it seemed) of Russians of all ages. This was a high-class performance, but we were especially impressed by the audience's enormous and enthusiastic response after the performance was over. Crowds surged toward the stage, the performers were showered with bouquets of flowers, and the clapping and shouts of approval went on and on.

We then made a special trip to Tbilisi, Georgia, which was then still part of the Soviet Union. But discontent over the political system was in the air. We sensed this especially when we were together with our hosts, a group of young mathematicians, all working in the area of category theory. This faithful and delightful group of category theorists took us sightseeing in Tbilisi to the best restaurants in town. There were flowers for Osa, new lectures to give and to hear, and splendid meals in keeping with the high standards of Georgian hospitality. We all went on a daylong excursion to the countryside where we explored one of the earliest Christian churches and visited the grave of the beloved Georgian poet Shot'ha Rusthaveli, who is compared with Shakespeare.

Later Developments

Saunders with Osa, holding her flowers, Republic of Georgia, 1987

Attendees of the Category Theory Gathering in Tbilisi, Georgia, 1987

Chapter Sixty — Second Marriage

It was an exceptionally effective exchange of ideas with this group of young Georgian mathematicians. In particular, George Janelidze was and is especially devoted and enthusiastic. I have kept in touch with him ever since. It is remarkable to see how abstract mathematical ideas have international resonance.

After Georgia we flew to Leningrad, where we again were well taken care of by my mathematical colleagues. For example, they arranged to get us inside the grand Hermitage Museum without having to wait in the seemingly endless line outside. They took us on tours through miserable prisons and luxurious, newly restored palaces outside Leningrad. We met a young algebraist, Soloviev, who had made decisive advances on my own earlier theorem stating that all (i.e., all suitable) diagrams commute.

From Leningrad we continued to Estonia, where I gave a talk at the Institute of Cybernetics in Tallinn and we were again greeted warmly by colleagues, both in Tallinn and at the University of Tartu. In Estonia too, we were much aware of the limits on freedom of speech. However, only a few weeks later glasnost and big changes took place in the Soviet system. Amazing! Within a few days, Georgians, Russians, Estonians, all were now allowed to communicate without fear. But this happened after we left.

On our way back home we stopped in Denmark where we again got a warm reception, this time by Osa's family, my new and lively relatives. Then on to Holland to spend a few days with Osa's sister Eva and her husband, Theo—ending with their taking us on lovely Dutch countryside trips.

All this during our first year of marriage!

Many trips have followed since, usually because of mathematical conferences and meetings, some near, some far away, adventurous, educational, and delightful—mathematically and otherwise. Mathematics has given us occasions to explore Spain, England, Italy, France, Portugal, Czechoslovakia, New Zealand, and a number of states within the United States. Once a year we take off for the meetings of the American Mathematical Society and the Academy of Sciences.

During the summer months at the Dunes we have many visits from our large combined family. Cynthia and Bill come from London.

Later Developments

Greta Schuessler, who was an associate at the Academy and has over the years become part of the family, comes from Washington. Gretchen turns up several times a year, sometimes with George. We also see sister-in-law Ingeborg, niece Alison, and their dog Beatrice several times a year.

Osa's children William, Andy, and Karen all camp with their families at the Dunes for a week during August, and in the springtime Osa's granddaughters spend their Easter vacation with us.

Much activity takes place when our visitors come: sewing, knitting, swimming, painting, reading, hiking, cooking, music, singing, tennis, and, of course, mathematics, which can at times be challenging when the house is occupied with campers. Indeed, my second marriage has presented a situation wholly different from my first.

Chapter Sixty One
International Category Conferences

Research in category theory and its applications has been encouraged by annual international conferences where experts from many countries convene to discuss current results and problems. For example, there was a splendid conference in 1990 at Lake Como in Italy that assembled experts from all over. There were stimulating talks, as well as some that were hard to understand (sometimes because too many words were crowded into transparencies and presented at high speed, a familiar problem). There was also a splendid cruise with dinner on Lake Como. The 1997 conference was held in Vancouver, Canada, with many fine talks.

The 1999 International Conference on Category Theory was held in Portugal, at the famous University of Coimbra; there, I presented my summary of Eilenberg's vital contribution to the development of category theory. The conference was organized by Professor Manuela Sobral, chairman of the mathematics department of Coimbra. There were over 150 people in attendance, representing at least 20 different countries. There were about 88 lectures on new results and categories, complete with some discussion and arguments (why not do it differently?). On Wednesday afternoon there were several elegant excursions—on one, we inspected an old hill-town fort. The conference dinner was held near Coimbra in an impressive castle that once belonged to the kings of Portugal and now belongs to the University. The entertainment included some traditional Portuguese singing. All told, it was a lively and productive session.

Later Developments

Saunders speaking at the International Conference of Category Theory, 1999

Saunders reading one of his poems

Chapter Sixty One ⁓ International Category Conferences

The Coimbra Conference had particular significance for me because the *Special Issue of the Journal of Pure and Applied Algebra*, in which the conference proceedings appeared, was dedicated to my 90th birthday. I prepared the following small poem about the meeting, which was read at the conference banquet.

Categories at Coimbra

Here's to the dual of each category
If you can't compute it you soon will be sorry
For duals and adjoints are everywhere found
And commutative diagrams soon will abound.

Sam Eilenberg said just one paper will do
To introduce categorical notions so new
We'll write it so well these ideas for to sell
And publish it promptly the story to tell.

Categories now flourish in north Portugal
We learn new ideas and constructions so swell
Coimbra's the scene of a categorist dream
Where all new results can appear in full gleam.

Separate mathematical specialties are often conducted too much in isolation. With this in mind, I once convinced the MSRI to arrange a joint conference on Universal Algebra and Category Theory, two subjects that could and might considerably overlap and help each other. The conference, held in 1993, had a number of good talks on such overlapping topics, but to my regret, it does not seem to have encouraged further interaction. For example, Lawvere's insightful description of a universal algebra is not used by the universal algebraists, and vice versa. Evidently, I have not yet found an effective way to encourage joint discussion and interchange.

Part Fifteen

Contemplating

Chapter Sixty Two
Mathematics Departments

Much of the stimulus for effective mathematical research comes from well-organized university departments of mathematics. This chapter will aim to summarize my own experience with several departments.

When I first became interested in mathematics in 1927 it was generally agreed that the three leading American departments of mathematics were those at Harvard, Chicago, and Princeton (an unordered list). Harvard combined its venerable traditions with a long-standing attention to mathematics, going back to the days (1870) of Benjamin Peirce, who was noted especially for his 19th-century research on linear algebra. From about 1900, William Fogg Osgood was the leading member of the Harvard department. He wrote the text for the Harvard calculus course (using infinitesimals instead of limits) and collaborated with his colleague W. C. Graustein in an undergraduate text on geometry. He was noted for his authoritative monograph on complex variable theory, written in German (his Ph.D. study had been in Germany). George Birkhoff assumed the leadership of the Harvard department, and had studied both at Harvard and at Chicago (with his Ph.D. from Chicago). As a young faculty member at Princeton he had proved Poincaré's last geometric theorem, which Poincaré had published without a proof. This accomplishment led to his recognition as the best American mathematician, and to his appointment at Harvard. There, with decisive results on differential equations, he soon became the leading

Contemplating

Deparment of Mathematics, University of Chicago, 1984

figure in the department; for example, he guided a number of notable Ph.D. students, including Marston Morse, Marshall Stone, B. J. Koopman, and D. V. Widder. George Birkhoff was devoted to mathematics and clearly disliked administrative tasks, although he did once serve for a period as dean of the Harvard faculty. He was sharply critical of mathematicians who gave up research for administration, and so he was careful not to do that himself. The Harvard department, in the period of my membership there, had a close and effective social structure with clear social cohesion. The total effect was an air of excitement about new mathematical results.

The Princeton Department of Mathematics had risen to eminence during Woodrow Wilson's presidency there, who brought in new faculty members such as Oswald Veblen (Ph.D. Chicago). Veblen and Dean Luther P. Eisenhart built up the department. Veblen's interest in logic led to the appointment of Alonzo Church, who soon had a number of leading students in logic, such as Stephen C. Kleene and Barkeley Rosser, as well as famous visitors such as Gödel and Turing. Veblen's interest in algebraic topology (as expressed in his influential book *Analysis Situs*) led to Princeton's appointment of Solomon Lefschetz, from Russia, who studied algebraic geometry

Chapter Sixty Two — Mathematics Departments

and topology. He was a dynamic person and kept life very lively, including his heckling of his more staid colleague Alonzo Church. There were many other active mathematics faculty members, such as Einar Hille in analysis and H. P. Robertson in relativity theory.

The Princeton mathematics department was housed in Fine Hall, named after the Princeton mathematician H. B. Fine, one of the founders of the American Mathematical Society. The Fine Hall coffee room collected faculty and graduate students for tea and talk, and games of chess and go. There many a theorem got started, by games or talk. The first-floor faculty offices were elegant, with open windows leading to the outdoors. Rumor has it that one distinguished professor, wishing to avoid a visitor knocking on his door, simply opened his window and stepped outside. This device is no longer available: the Princeton mathematics department is housed in a new Fine Hall—a high-rise building not open to this escape maneuver, but much mathematics is still carried on in the tearoom there.

The University of Chicago department started with the founding of the university in 1892 under President William Raney Harper. Harper wanted a research university, so he brought many distinguished scholars to the initial Chicago faculty. As chairman of mathematics he chose a young Ph.D. from Yale, Eliakim Hastings Moore. Then Moore and Harper managed to attract (from Clark University) two distinguished German mathematicians, Oscar Bolza, an expert in the calculus of variations, and Heinrich Maschke, an algebraist. The new department immediately emphasized graduate study; soon the Midwest departments of mathematics (Iowa, Michigan, Wisconsin, etc.) had as chairmen Ph.D.s from Chicago. But new fields can develop unexpectedly and older specialties can become worn-out subjects. At Chicago, the inheritance method of the Bliss department was not effective. Mathematics greatly needs, from time to time, new thoughts and new directions of critical study.

My lengthy experience at Chicago began when Marshall Stone came from Harvard to Chicago in 1947 to wholly reorganize and rebuild the department there. With the support of President Robert M. Hutchins, he quickly established a new and vital department, with an international cast: Chern from China, Zygmund from

Poland, Weil from France, and Adrian Albert from Chicago. This group established a wholly new graduate program, with a sequence of newly designed courses (Point Set Topology, Analysis, Algebra, etc.). A great group of graduate students came, many of whom were supported by the GI Bill. The mix was a dynamic interaction between professors and lively graduate students. For my part, I wrote up lecture notes on topology, on algebraic topology, and on aspects of algebra, as well as some notes on axiomatic set theory.

The whole atmosphere during the Stone Age was electric, with long arguments at tea about mathematics. I have presented a summary in my article "Mathematics at the University of Chicago: A Brief History," published in the 1992 AMS issue, *A Century of Mathematics in America*. It was a great time, though it did not last, and in a sense it gradually came to an end in the late 50s. But while it lasted, it was exciting and full of new mathematics.

At one point, the University of Chicago set up a faculty committee, chaired by Professor Edward Shils, to examine and report on the Criteria for Academic Appointment. I was one of the members of this committee. We duly, and I believe vigorously, so reported. For example, "The Committee regards distinction in research accomplishment and promise as the sine qua non of academic appointment. Even where a candidate offers promise of becoming a classroom teacher of outstanding merit, evidence should be sought as to the promise of distinction in his research capacity. Even if his research production is small in amount, no compromise should be made regarding the quality of research done... It is imperative that in every case the appointing body ask itself whether the candidate proposed, if young, is likely in a decade to be among the most distinguished scientists or scholars of his generation...." In my opinion the work of the committee set useful standards for developing and maintaining excellent mathematics departments.

The trio of mathematics departments, Princeton, Harvard, and Chicago, by their activity, did set the style and direction for American mathematics in the period 1892-1930. Other departments developed and competed, but it was still the case that much of the most exciting mathematics went on at Princeton, Harvard, and Chicago. By now there

Chapter Sixty Two — Mathematics Departments

are many other American universities active in mathematical research. But even the arrangement of a leading three institutions had an advantage over the European scene in mathematics. There, English mathematics had long been dominated by Cambridge (with an assist from Oxford). In France, Paris is the unique center of everything, including mathematics. For example, at some periods it was common for a professor at a provincial university to live in Paris and commute weekly to his teaching obligations. The result has been a lively and stimulating center of research in Paris, far ahead of the provincial universities.

Germany, with a divided political past, has a much more varied university system. But in mathematics, there is a traditional leadership in Göttingen and Berlin. The Göttingen tradition goes back at least to Gauss and his student Riemann; subsequently, Felix Klein managed the department at Göttingen with great finesse. His own mathematical research career had ended early, but his talents as an organizer and leader were impressive and effective. He had a vital interest in the improvement of teaching, as in his well-known book *Elementary Mathematics From a Higher Standpoint*. In Göttingen the sense of tradition was strong. Along the city wall near the Mathematical Institute the famous monument with Gauss and Weber attested to this tradition.

Berlin gloried in its tradition, such as the widely circulated 19th-century lectures of Weierstrass, which had codified the rigorous epsilon-delta foundation for the calculus. There was also in Berlin that pioneering emphasis by Frobenius and Schur on group theory. Berlin competed with Göttingen. At one point in the 1950s, I visited Berlin, seeing both the American and the Eastern Russian Zones. There, one of the Berlin mathematicians told me of the files of old proposed appointments. Berlin had considered trying to hire Hilbert away from Göttingen, but one of the letters of support expressed doubts about Hilbert: He had contributed to algebraic number theory in his famous report on the state of that subject, but with this he had perhaps stopped research work. At any rate, Hilbert stayed in Göttingen, continued his deep research and influenced many students.

There were other active mathematicians in Germany when I was studying there. I was especially aware of Munich, where Caratheodory was active in the calculus of variations and the theory of point-set measure at the university there. Algebraic topology was a topic of activity in Heidelberg, and in Hamburg the new university had an especially lively department, with Blaschke in differential geometry and Emil Artin in algebra and class field theory. All told, Germany presented (before the war) a splendid and lively mathematical creativity.

In earlier times, many of these German departments had provided ideas and stimuli to young American mathematicians studying for the Ph.D. In the United States the ideas of active departments of mathematics have been well and vigorously developed. The German model evidently had a considerable influence on the department at Chicago, both in its earlier stages under Eliakim Moore and during the decisive Stone Age.

Building up and maintaining a strong department of mathematics requires consummate care in the making of appointments. There is a constant temptation for more of the same, that is, appointing new members in specialties already represented. But it is vital to resist this temptation.

Personally, I have been privileged to have had direct contact with three remarkable departments of mathematics: Göttingen before the Nazis, Harvard under the guidance of George David Birkhoff, and Chicago in the Stone Age.

Chapter Sixty Three
Collaborative Research

The opportunity to collaborate in writing mathematics and in research has been exceptionally important for me. First, there was the happy circumstance that Garrett Birkhoff and I were both active junior faculty members at Harvard, and both working in algebra at the time when the modern (Emmy Noether/van der Waerden) approach to algebra was just becoming available. We jointly seized on this situation to codify our teaching of the subject, resulting in our joint undergraduate text *A Survey of Modern Algebra*. We designed the book to present both the concrete examples of groups, rings, and fields and their axiomatic treatment. It was published at the right time (1942), just as it was becoming clear that this was the necessary way to present this subject matter for undergraduate courses in algebra. We were both pleased with its initial success and its subsequent popularity. Garrett and I did not collaborate then on our actual research projects, but we did collaborate on a shorter version of *Survey* and on a later, more categorically abstract text *Algebra*. The high point was clearly *Survey*, written when we were young and inclined to lead a mild revolution in the way undergraduates were taught.

Thus, drastic changes may be developed well through collaborative work. At any rate, that was the case not just for *Survey*, but for Moerdijk and me on topos theory as well. There, the novelty of the subject very much required our joint efforts and our combined knowledge.

Projects done in collaboration can be especially effective when they combine different aspects of knowledge; it can help particularly when mathematics is changing. Thus, on visits with Dorothy to Arkansas I collaborated with Professor V. W. Adkisson there; his knowledge of the intricacies of point set topology combined with my more limited experience in that field. His interests presented me with specific problems that I found fascinating, and that we were largely able to settle together. When the problems were solved, our collaboration came to a natural end.

My collaboration with Otto Schilling on algebra and in class field theory were inspired by our (fleeting) early acquaintance in Germany, by Otto's vast knowledge of class field theory, and by my insistent curiosity about the subject, which I had failed to understand when I listened to lectures in Göttingen by Emil Artin. My second joint paper with Schilling established useful properties of algebras, and it was from them that I learned about crossed product algebra and factor sets. For me, collaboration was a very effective way to learn new subjects in the company of experts. I regret the fact that I was later (during the war) not able to help Otto with the writing of his big book on valuations.

I also had other shorter, but effective, collaborations with my Yale almost-classmate Alfred H. Clifford on matters of common interest in group theory; with Miriam Becker on the minimal number of generators for inseparable algebraic field extensions; and with F. K. Schmidt (in Heidelberg) on such questions in answer to my discovery of an error in a paper by Hasse and Schmidt. My work with Eilenberg on topology led naturally to a joint paper with Henry Whitehead on the use of the cohomology of groups to determine homotopy type, that determination being a long-term interest of Whitehead's. Whitehead was an exciting and dynamic collaborator. He once quite properly chided me for inactivity and provided a sketch of a lazy Mac Lane.

All these examples of collaboration must yield in size and consequence to my long continued joint work with Eilenberg. He and I came together on a problem that combined our expert knowledge in topology and algebra. The combination turned out to be very fruitful, leading to our many major joint papers and covering discoveries such as the cohomology of groups, homological algebra, Eilenberg-Mac Lane spaces, and category theory, as previously discussed. It was the interaction between solenoids and group extension that got our collaboration started, and this first work of

Chapter Sixty Three ~ Collaborative Research

collaboration revealed much else to be done, some stimulated by a result of Heinz Hopf. It can be best summarized by a striking proof that 2 is more than 1 plus 1.

This collaboration is a relevant example of the reception in the United States of European refugee mathematicians. When Sammy got out of Poland in 1939, he found a position at the University of Michigan. And it was there that Sammy and I started our collaboration, one small, but typical, example of East meets West.

The great influx of refugee mathematicians from Europe presented a decisive stimulus for American mathematics in the 1940s. Eilenberg was one of many displaced Europeans who came here, were guided to university positions, and stimulated research by their collaborations and their publications. Veblen, Lefschetz, and others at Princeton deserve credit for their effective efforts to get suitable positions for refugee mathematicians.

Chapter Sixty Four
Career Choice: Inheritance of Precision?

My family background emphasized the ministerial element: My father was a devoted Congregational minister, as was his father, W. W. McLane, though W.W.M. had started out as a Presbyterian minister, in keeping with Scottish tradition. Various other ancestors had also followed a ministerial life, for example, a second cousin, Ami Ruhami Robbins. It might have been natural and easy for me to choose for myself such a career, but I did not do this. One cause may be that my father fell ill before I ventured on a choice. I might have discussed such a choice with my grandfather McLane, but the gap of two generations would have made this harder. For instance, I once asked him about the purpose of life; he responded that we lived for the glory of God, but I did not really understand what this meant. I also recall that I had various theological doubts and questions. When I officially joined my grandfather's Congregational Church in Leominster, I had to answer a long list of questions; I recall that I did answer suitably but, on each question, kept in mind a rationalized qualification. Moreover, by that time I was fascinated with high-school chemistry and I had more or less decided that I would like to become a chemist. This, I thought, would be interesting work that would probably pay me an adequate salary—I distinctly felt that my father's salary had been not quite adequate. As I mentioned earlier, I calculated that if I saved enough income, I would reach retirement with a capital of about $100,000, and that the interest at 4 percent would be sufficient for my old age. Of course, this figure turned out to be wildly wrong, but at the time I had little understanding of the working of the economy (I had not yet heard of socialism, and I did not anticipate inflation).

The combination of these ideas meant that I started college with the aim of finding a career that would be scientific rather than ministerial. By this time my father had died. I treasured his letters, but they were concerned more with the general aspects of life rather than with the choice of career. However, the effects of the ministerial background were actively present. I had seen my father struggle with the draft of his sermons, perhaps not so different from the draft of scientific papers. And teaching resembled preaching, since the student's mind is like his soul—both need repeated refreshment. A minister must write and speak clearly, as must a teacher. But a ministerial career would have required my belief of things of which I was uncertain. Mathematics, however, provided a different sort of certainty, but that is a comparison I make only hesitantly.

Inheritance is probably not the major source and structure of my activity in mathematics, although my Uncle John did make it possible for me to continue a family tradition and attend Yale as an undergraduate. There, Lester Hill, my freshman mathematics instructor, gave me the idea of a possible career in mathematics. This was reinforced by other teachers: Wallace Wilson, who introduced me to topology by way of the Hausdorff book; Oystein Ore, whose lectures revealed to me the new excitement in Emmy Noether's algebra; and Egbert Miles, who taught me advanced calculus and epsilontics, and who also showed me how to write mathematics. These Yale faculty members in mathematics and F.S.C. Northrop in philosophy made my undergraduate work inspiring. I was indeed fortunate to study at Yale.

Then later came the joys and achievements of collaboration with Garrett Birkhoff, Samuel Eilenberg, and others. I have been fortunate in the way that my early preparation helped me to understand and to express both the precision and the structure of mathematics. The resulting collaborations were clearly the high point of my research activity in mathematics. The most surprising aspect was the way that precision came to play in my later activity for the Report Review Committee of the National Academy of Science—discussion of science policy needs precision.

All told, mathematics was a great career choice for me.

Chapter Sixty Four ~ Career Choice: Inheritence of Precision?

Attendees of a conference at Columbia honoring Saunders' 90th birthday, 2000

Index

A

Abraham, Ralph 243
Adams, Frank 246, 249, 313
Adkisson, Virgil W. 73, 93, 346
Adler, Mortimer J. 40, 184
Ahlfors, Lars 69, 135
Alaoglu, Leonidas 135, 136
Albert, A. Adrian 79, 96, 171, 172, 175, 180, 181, 219, 342
Alberts, Bruce 228
Alperin, Jon xii
Andrews, Isabel 5, 117, 118
Angell, James Rowland 34, 141, 183
Applegate, H. 239
Artin, Emil 50, 79, 94, 344, 346
Askey, Richard xi
Assmus, E. F., Jr. 239
Awodey, Steve 314
Ayres, W. L. 72

B

Baer, Reinhold 97, 99, 175
Baily, Walter 181
Ballard, William 312
Barnard, R. W. 39
Barr, Michael 239, 240
Bass, Hyman 239
Beams, Jesse W. 24
Beatley, Ralph 206
Beck, Jon 239, 240
Becker, Miriam 346
Beissinger, Janet xv
Benabou, Jean 240, 241
Benedict, Manson 40

Bernays, Paul 45, 47, 51, 52, 53, 54, 60, 162
Bessel-Hagen, Erich 46
Birkhoff, Garrett 50, 72, 80, 81, 82, 136, 137, 162, 227, 345, 350
Birkhoff, George David 36, 38, 69, 70, 84, 206, 339, 340, 344
Blaschke, Wilhelm 171, 344
Bliss, Gilbert Ames 36, 37, 38, 78, 167
Bloch, Erich 283
Boas, Ralph 202
Bolza, Oscar 37, 341
Borel, Armand 156
Bourbaki, Nicholas 201, 202, 203, 205
Bradford, William 4
Brady, Geraldine 315
Brauer, Richard 87, 97, 175
Bronk, Detlev 227, 228, 229, 230, 234
Browder, Earl 178
Browder, Felix xii, 178, 179, 288
Brown, E. W. 31, 34, 115, 243
Buchi, Richard 161
Buchsbaum, David 209
Bush, Merrill 16
Bush, Vannevar 279, 322
Butler, Nicholas 141

C

Calderon, Alberto 174
Calinger, Ronald 315
Callahan, Jean 305
Caratheodory, Constantin 343
Cartan, Elie 171
Cartan, Henri 132, 156, 219, 221, 223, 237, 315

Index

Carter, W. C., Jr. 88
Cauer, Richard 48, 159
Cech, Eduardo 127, 128
Chandrasekhar, Subramanyan 243
Charlap, Leonard 239
Chern, S. S. 171, 174, 181, 243, 341
Christy, Dan 249
Church, Alonzo 76, 198, 199, 340, 341
Clifford, Alfred H. 98, 346
Cohen, Leo 121
Cohen, Paul 309, 311
Conant, James B. 137, 141, 227, 228
Cooley, John C. 24, 144
Coolidge, Julian Lowell 69, 70
Courant, Richard 48, 53, 114
Curry, Haskel 311
Curtis, John 74

D

Debus, Alan 315
Decker, Gerald John 220, 315
Dickson, Leonard E. 37, 79, 96, 97, 171
Dold, Abrecht 238
Dole, Harry 313
Dubreil, Paul 155, 221
Dubuc, Eduardo 244, 313
Dyer, Eldon 181, 189

E

Eckmann, Beno 129, 157, 240
Edison, Thomas 230
Edsell, John 234
Edwards, Kathleen 313
Eilenberg, Samuel xi, xii, 73, 88, 93, 98, 100, 102, 103, 104, 106, 109, 110, 120, 125, 126, 129, 131, 132, 157, 162, 163, 168, 181, 191, 192, 209, 210, 227, 233, 237, 239, 240, 249, 250, 251, 333, 335, 346, 347, 350
Eisenbud, David xvi, 312
Eisenhart, Luther P. 340
En-Lai, Chou 294
Euler, Leonard 191, 315
Everett, H. S. 172

F

Feit, Walter 175
Feldman, Chester 239
Fine, H. B. 341
Fingerman, Joel 315
Franz, Wolfgang 159, 215
Freudenthal, Hans 129
Freyd, Peter 202, 239, 240
Friedrichs, Kurt 114
Frobenius, Georg 343
Frohlich, A. 249

G

Gehman, Harry 198
Geiger, Moritz 55
Gelfond, A. O. 139
Gentzen, Gerhard 49
Gibbs, Josiah Willard 4, 115, 233
Ginali, Suzanne 314
Gödel, Kurt 51, 340
Gove, Wallace 15
Grad, Arthur 280
Graustein, William C. 67, 68, 69, 339
Graves, Lawrence 172
Gray, John 239, 240
Greensfelder, Olive 16, 18, 77, 82
Grothendieck, Alexander 174, 209, 210
Grubb, Bill 15
Grubb, John 15
Guggenheim, Victor 213

H

Hadamard, J. 174
Hall, Marshall 97, 175
Hall, Philip 82, 162
Halmos, Paul xii, 172, 180
Halpern, Edward 312
Hamilton, W. R. 94
Hamsher, Ross 220, 315
Handler, Philip 228
Hardy, G. H. 46, 47
Harper, William Raney 341
Harrington, Marjorie 28, 77, 117, 118, 143, 245

Hasse, Helmut 83, 84, 93, 94, 97, 158
Hay, Cynthia. See Mac Lane, Cynthia
Hay, William 299, 300, 301, 331
Heidegger, Martin 161
Helgason, Sigurdur 181
Herglotz, Gustav 48, 54, 55, 79, 158
Herstein, I. N. 172
Hestenes, Magnus 69, 74, 78, 120
Hilbert, David 44, 45, 47, 50, 94, 343
Hildebrandt, Theophil H. 36, 99
Hill, George William 115
Hill, Lester 22, 350
Hille, Einar 341
Hilton, Peter 240
Hirsch, Guy 162
Hirzebruch, Fritz 180
Hitler, Adolf 52, 53, 54, 57
Hochschild, Gerhard 89, 237
Hodge, W. V. D. 179
Hopf, Heinz 128, 129, 157, 158, 347
Hopper, Grace Murray 36
Hotelling, Harold 121
Howard, William 310–311, 315
Hua, L. K. 293, 294
Hungerford, Thomas 312
Hunt, Pearson 23
Huntington, E. V. 69, 115
Hutchins, Robert Maynard 35, 36, 40, 41, 77, 80, 141, 167, 169, 178, 183, 184, 185, 186, 187, 189, 341

J

Jacobson, Nathan 50
James, Ioan 213
Janelidze, George 331
John, Fritz 49, 56, 114
Johnstone, Peter 313
Jones, Alice 73, 301
Jones, Dorothy xii, 40, 52, 56, 65, 68, 73, 74, 77, 79, 81, 84, 117, 136, 147–154, 157, 158, 160, 161, 162, 167, 169, 177, 185, 186, 189, 211, 212, 213, 215, 216, 219, 220, 221, 238, 245, 246, 295–296, 299, 300, 301, 303–306, 313, 327, 328, 345
Jones, Isabel 73, 301
Jones, Virgil L. 73

K

Kadison, Richard 175
Kan, Daniel 191, 239, 250
Kaplansky, Irving x, xiii, xvi, 87, 88, 120, 172, 237
Kegel, Otto xii
Kelly, Max 240
Kimpton, Lawrence A. 178, 180, 185, 187, 206, 207
Kirkpatrick, Annis 16
Kirkpatrick, Ralph 15, 16
Kleene, Stephen C. 73, 76, 138, 198, 340
Klein, Felix 47, 48, 83, 343
Kline, J. R. 73
Knopp, Konrad 215
Kock, Anders 244
Kohn, Joseph 287
Koopman, B. J. 340
Kostant, Bert 173, 175
Krasner, Marc 155, 221
Kristensen, Leif 312
Krull, Wolfgang 66
Kruse, Arthur 311
Kuo, T. C. 312
Kuratowski, C. 72
Kurosh, A. G. 328

L

Landau, Edmund 37, 46, 47, 53, 59
Lane, Ernest P. 39, 167, 294
Lang, Serge 179, 180, 181
Lashof, Richard 189
Latimer, Wendell 227, 228
Lawrence, Ernest O. 25, 32
Lawvere, F. William 191, 192, 193, 239, 244, 250, 313
Lax, Peter 114
Lefschetz, Solomon 76, 101, 102, 250, 340, 347
Leighton, Walter 120

Index

Leray, Jean 88, 156, 157
Levi, Edward 141
Lewis, Daniel 74, 121
Lewy, Hans 48, 54
Liao, S. D. 296
Linton, Fred 239
Liulevicius, Arunas 312
Longley, William Raymond 23
Lorenz, Edward 89
Lowell, A. L. 72
Lubkin, Saul 239
Lunn, A. C. 39
Luogeng, Hua. See Hua, L. K.
Lyndon, Roger 88, 89, 157

M

MacDonald, John xvi, 238
Mackey, George 125, 135
MacLane, David 11, 12, 13, 17, 28, 144, 306
MacLane, Donald Bradford 3, 5, 6, 7, 8, 10, 12, 17, 349, 350
MacLane, Duncan 144
MacLane, Gerald 9, 11, 13, 17, 28, 144
MacLane, John F. 3, 17, 18, 25, 26, 142, 350
MacLane, Lois 7, 8
MacLane, Paul 3, 17
MacLane, Stanley 17
MacLane, William 17
MacLane, Winifred. See Saunders, Winifred
MacLean, Charles Hector Fitzroy 213
Mac Lane, Cynthia xvi, 117, 118, 136, 143, 155, 167, 211, 212, 213, 214, 215, 217, 220, 222, 299, 301, 303, 306, 331
Mac Lane, Dorothy 6. See Jones, Dorothy
Mac Lane, Gretchen xvi, 7, 77, 117, 118, 136, 155, 167, 211, 212, 213, 214, 215, 217, 220, 221, 299, 301, 303, 313, 332
Mac Lane, Osa. See Skotting, Osa

Magnus, Wilhelm 158
Maschke, Heinrich 341
Mason, Max 116
May, Maya 300, 305
McKenzie, Bob 23
McKeon, Richard 314–315
McLane, Frances 13
McLane, John 3
McLane, Julia 3
McLane, William Ward 3, 4, 13, 16, 17, 28, 349
McShane, E. J. 279
Menger, Karl 80
Meyer, Herman 189
Miles, Egbert J. 26, 27, 35, 65, 82, 350
Mitchell, William 239
Moerdijk, Ieke 314, 345
Moore, Eliakim Hastings 36, 37, 38, 39, 116, 192, 197, 316, 341, 344
Moore, John 240
Morley, Michael 310
Morrison, Harvey 21
Morse, Marston 38, 69, 120, 279, 340
Mostert, Paul 249
Moulton, F. R. 116
Moyls, Benjamin N. 89, 312

N

Neisendorfer, Joe xii
Nerode, Anil xiii, 310, 315
Neugebauer, Otto 48, 54
Newcomb, Simon 115
Noether, Emmy 32, 33, 45, 47, 48, 49, 50, 53, 66, 82, 84, 97, 98, 210, 314, 350
Noether, Max 33
Northrop, Eugene 32, 189
Northrop, Filmer S. C. 23, 24, 144, 350
Nunke, Ronald 312

O

Ore, Oystein 32, 33, 35, 36, 40, 65, 66, 67, 80, 82, 198, 350
Orleans, Leo H. 295

Index

Osgood, William Fogg 69, 136, 339
Ostrowsky, A. 158

P

Paige, Leigh 33, 34
Palmquist, Paul xvi, 313
Pasta, Bill 287
Pauling, Linus 234
Peirce, Benjamin 339
Peirce, Charles Sanders 315
Pell, William 287
Phelps, Russell 87
Phelps, William Lyon 31
Pierpont, James 24, 31, 36
Polya, George 44
Porter, R. J. 239
Postnikov, M. 163
Prager, William 113
Preobrajenskaya, Olga 221
Press, Frank 228
Puppe, Dieter 238
Putnam, Alfred 87, 189

Q

Quine, Willard Van Orman 72, 145

R

Rainville, Earl D. 14
Randolph, John 74
Reidemeister, Kurt 215
Rellich, Franz 48, 158, 215
Richards, Alfred N. 227
Richardson, R. G. D. 113
Ripley, Julian D. 24, 25
Robbins, Ami Ruhami 18, 349
Robertson, H. P. 116, 341
Robinson, Frances 4
Robinson, Julia 258
Robson, Chris xiv
Rosen, William 287
Rosser, Barkeley 73, 74, 76, 138, 340
Rosser, J. B. 198
Rudall, Nicolas 305
Russell, Bertrand 138, 139

S

Sard, Arthur 120
Saunders, Aretas 5
Saunders, Aretas, Jr. 6
Saunders, Dorothea 6
Saunders, George Aretas 5
Saunders, Winifred 5, 6, 12, 13, 21, 28, 59, 77, 192, 245
Schilling, Otto 93, 97, 99, 172, 346
Schmidt, Arnold 158, 215
Schmidt, Bill 10
Schmidt, F. K. 83, 84, 346
Schreiber, C. F. 23
Schreier, Otto 97
Schubert, Horst 160
Schuessler, Greta xvi, 301, 332
Schuman, Fred 186
Schur, Issai 343
Scott, Dana 309, 310
Segal, Andrew 327, 328, 332
Segal, Irving 172, 223, 327
Segal, Karen 327, 328, 332
Segal, Osa. See Skotting, Osa
Segal, William 327, 328, 332
Seifert, H. 160
Seitz, Frederick 228, 230
Serre, J. P. 89, 156, 210
Seymour, Charles 142
Shafer, David 311
Shils, Edward 342
Shukla, U. 222
Siegel, Carl Ludwig 46
Siegler, Mark 303
Singer, Isadore 174, 175
Skolem, T. 315
Skotting, Doetter 327
Skotting, Eva 327, 331
Skotting, Lasse 327
Skotting, Osa xvi, 223, 327, 328, 329, 330, 331
Skotting, Poul 327
Slaught, Herbert E. 197
Smith, H. L. 316
Smith, Paul 120, 125
Snyder, H. B. 75

Solovay, Robert 309
Soloviev, A. 313, 331
Spanier, Edwin 172, 174, 181
Speiser, A. 158
Springer, Ferdinand 159
Stanley, Wendell 233, 234
Stauffer, Howard 316
Steenrod, Norman 93, 101
Stein, Elias 174, 287
Sternberg, Shlomo 181
Stigler, Steve 305
Stoker, James J. 114
Stone, Emmy 327
Stone, Harlan F. 169
Stone, Marshall 69, 70, 80, 125, 135, 136, 137, 167, 168, 169, 171, 172, 173, 177, 179, 181, 311, 340, 341
Su, Buqing 294, 296
Szczarba, Robert xiii, 312

$\mathcal{T}$

Teichmüller, Oswald 48, 49, 97
Thaler, Alvin 287
Thompson, John x, xiii, 175, 312
Thoms, Rene 288
Threllfall, W. 160
Tierney, Myles 239, 313
Tinker, Chauncy Brewster 32
Truesdell, Clifford 191
Truman, Harry S. 283
Tsun, Wu Wen 296
Tukey, John 114, 115
Turing, Alan 97, 340

$\mathcal{V}$

van der Waerden, B. L. 82
van Vleck, J. H. 115, 116
Veblen, Oswald 36, 101, 162, 198, 340, 347
von Mises, Ludwig 50
von Neumann, John 49, 50, 70, 167, 172, 179
Voreadou, Rosie 313

$\mathcal{W}$

Walker, Robert 74, 75, 76, 77
Wallis, Allen 121
Walsh, Joseph 69, 135, 138
Waterman, Alan 283
Weaver, Warren 116, 119, 120
Wedderburn, Joseph H. M. 96
Weierstrass, Karl 343
Weil, André 85, 135, 168, 171, 173, 174, 175, 178, 180, 202, 209, 310, 311, 342
Weyl, Hermann 44, 45, 50, 51, 52, 54, 55, 59, 61, 83, 114
Whitehead, Alfred North 163
Whitehead, J. Henry C. 162, 163, 213, 346
Whitney, Hassler 70, 71, 72, 73, 76, 93, 120, 162, 168, 169
Wicht, M. C. 316
Widder, David V. 69, 340
Wielandt, Helmut 215
Wiener, Norbert 139
Wilczynski, E. 294
Wilder, R. L. 93
Wilks, Samuel 114, 115
Wilson, Edwin B. 115, 233, 234
Wilson, Wallace A. 23, 26, 73, 80, 350
Wilson, Woodrow 8, 229, 230, 340
Witt, Ernst 48, 49
Wittgenstein, Ludwig 325
Wolcott, Helen 16
Wright, Donald 23, 30

$\mathcal{Y}$

Yao, Joseph Z. T. 312

$\mathcal{Z}$

Zahler, Raphael 243
Zariski, Oscar 67, 76
Zermelo, E. 36, 37, 192
Zervos, P. 313
Zilber, J. 251
Zygmund, Antoni 171, 172, 173, 174, 341